Energy
Management

PRENTICE-HALL SERIES IN ENERGY

Wayne C. Turner and W. J. Kennedy, Jr., editors

Energy Management

W. J. Kennedy, Jr.
University of Utah

Wayne C. Turner
Oklahoma State University

Prentice-Hall, Inc., Englewood Cliffs, N.J. 07632

Library of Congress Cataloging in Publication Data

KENNEDY, WILLIAM J.   (date)
    Energy management.

(Prentice-Hall series in energy)
    Bibliography: p.
    Includes index.
    1. Industry—Energy conservation.   I. Turner, Wayne C.
(date).   II. Title.   III. Series.
TJ163.3.K45   1984   658.2'6   83–22970
ISBN   0-13-277674-X

Editorial/production supervision
    and chapter opening design: *Gretchen K. Chenenko*
Cover design: *Edsal Enterprises*
Jacket design: *Ben Santora*
Manufacturing buyer: *Anthony Caruso*

Printed in the United States of America

10 9 8 7 6 5 4 3 2 1

ISBN   0-13-277674-X

PRENTICE-HALL INTERNATIONAL, INC., *London*
PRENTICE-HALL OF AUSTRALIA PTY. LIMITED, *Sydney*
EDITORA PRENTICE-HALL DO BRASIL, LTDA., *Rio de Janeiro*
PRENTICE-HALL CANADA INC., *Toronto*
PRENTICE-HALL OF INDIA PRIVATE LIMITED, *New Delhi*
PRENTICE-HALL OF JAPAN, INC., *Tokyo*
PRENTICE-HALL OF SOUTHEAST ASIA PTE. LTD., *Singapore*
WHITEHALL BOOKS LIMITED, *Wellington, New Zealand*

Contents

v

. . . .

The Energy Audit   29

. . . .

Energy Sources and Rate Structures   53

Chapter 9

· ·

Control Systems and Computers   177

. . . .

. . . .

. . . .

Preface

This book is a culmination of many years of experience in energy management and is a result of our dedication to the concept of prudent energy utilization for maximum return per unit of energy consumed. We have seen organizations achieve phenomenal results in energy management by applying sound engineering principles and utilizing creativity. Savings in energy of 30% are usual. Savings often run 40–50%, and a 70% saving, though uncommon, has been observed several times. Since these figures grow larger as technology matures, the potential is exciting and motivating.

The purpose of this book is to help you understand the objectives of energy management and to demonstrate and teach some of the basic techniques and tools. This book is designed to be a text book aimed at juniors, seniors, and first-year graduate students. The text is self-contained in that no other texts should be required for a course in energy management. Practitioners should find this book a useful addition to their libraries, but again, the primary objective is use as a text in a college course or in continuing education short courses.

The book is organized as the material would be taught in a college course. In the first chapter, energy management is defined in depth. Then the proper tools for designing, initiating, and managing energy management programs are presented. In Chapter 2 we discuss energy auditing. Suggested forms are provided. Chapter 3 is a presentation of utility rate structures for various types of energy. Emphasis is on understanding and interpreting the various structures. Engineering economy and the proper treatment of inflation are given top priority. Chapters 5–10 are treatments of energy management applications in various areas of the organization, including service functions such as maintenance and steam distribution. Chapter 11 is a discussion of insulation. In Chapter 12 we show how to analyze industrial processes, with emphasis on applications that are now or soon likely to be cost effective. Finally, in Chapter 13 we discuss alternative energy sources and water management.

We suggest that the book is of nearly ideal length for a one-semester course in energy management and that the instructor should start at Chapter 1 and proceed through Chapter 13 in order (at least for the first time—after that, the instructor could easily "jump around" as he or she wishes). Some chapters will probably be skipped or covered lightly, while most can be supplemented with the references provided. For example, if this course is taught in an industrial engineering curriculum, the material in Chapter 4 (economic analysis) could be covered very quickly; mechanical engineers might cover Chapter 6 (environmental control) quickly, etc.

As with most books, we could dedicate this to many people and ideals that have helped shape our lives and thoughts. Our wives (Anna and Kathy) and our children suffered most from time demands. Without them, this effort would not be worthwhile. Finally, we both believe strongly in free enterprise, the American way, and protection of our wondrous environment. This book is dedicated to all of these in the hopes that each will be a little better off because of it.

W. J. K., Jr.
W. C. T.

Chapter 1

. . . .

Designing, Initiating, and Managing an Energy Management System

1.0 ENERGY MANAGEMENT

The words *energy management* mean different things to different people, so perhaps the first objective is to define the term. To us, energy management is

> The judicious and effective use of energy to maximize profits (minimize costs) and enhance competitive positions

Therefore, any management activity that affects the use of energy falls under the overall guise of energy management. This is a rather broad intrepretation as it covers many areas of an enterprise's activities from product and equipment design through product shipment. It also must include waste disposal activities as many energy management opportunities can occur there. Without this *whole systems* viewpoint, many important activities will be overlooked or suboptimized.

We've already stated that the primary objective is to maximize profits or minimize costs. Some alternative statements or subobjectives are the following:

1. To conserve energy, thereby reducing costs
2. To cultivate good communications on energy matters
3. To develop and maintain effective monitoring, reporting, and management strategies for wise energy usage
4. To do research and development to find new and better ways to increase returns from energy investments
5. To develop interest in and dedication to the energy management program from all employees

1

6. To reduce the impacts of curtailments, brownouts, or any interruption in energy supplies

This list could continue, but these six seem to be of primary importance and enough for the purposes here. However, we should elaborate somewhat on objective 6. Curtailments occur when a major supplier of an energy source is forced to reduce shipments or allocations (sometimes drastically) because of severe weather conditions and/or distribution problems. This happens in the case of natural gas where industry is given the advantage of relatively inexpensive gas but on an interruptible basis. That is, residential customers and others on noninterruptible schedules have priority, and those on interruptible schedules receive what is left over. This residual is normally sufficient, but periodically curtailments of gas deliveries are necessary.

Even though they don't occur frequently, the cost associated with curtailments is so high (sometimes complete shutdown is necessary) that management needs to be alert in order to minimize the negative effects. There are several ways of doing this, but the one most often employed is the storage and use of a secondary or standby fuel. No. 2 fuel oil is often stored on site and used in boilers capable of burning either natural gas (primary fuel) or fuel oil (secondary fuel) when curtailments are imposed. Naturally, the cost of equipping boilers with dual fire capability is high, as is the cost of storing the fuel oil. These costs, however, are miniscule compared to the cost of forced shutdown. Other methods of planning for curtailments include production scheduling to build up inventories for curtailment-likely periods of time, planned plant shutdowns, or vacations during curtailment-likely periods, and contingency plans whereby certain equipment, departments, etc., will be shut down so critical areas can keep operating. All these activities are part of energy management.

Please note from the preceding discussion that energy conservation is an important part of energy management but that there is much more to energy management than just conservation. Curtailment-contingency planning is certainly not conservation, and neither is load shedding or power factor improvement in electrical energy management (to be discussed). To concentrate solely on conservation would preclude some of the most important activities—often those with the largest savings opportunity.

1.1 WHY?

1.1.1 Economics

The American free enterprise system operates on the necessity of profits (budget allocations in the case of nonprofit organizations), so any new activity can be justified only if it is cost effective; that is, the net result must show a profit improvement or cost reduction greater than the cost of the activity. Energy management has proven time and time again that it is cost effective.

Energy cost savings of 5–15% are usually obtained quickly with little to no required capital expenditure when an aggressive energy management program is launched. Eventual savings of 30% are common and savings of 50, 60, and even 70% have been obtained. All this is for retrofit activities. New buildings designed to be energy efficient often operate on 20% of the energy normally required (80% savings). In fact, we can boldly state that for most manufacturing and other commercial organizations *energy management is one of the most promising profit improvement-cost reduction programs available today.*

1.1.2 National Good

There are other reasons energy management programs are so vitally needed today. One of the most important is that energy management helps the nation face some of its biggest problems. Perhaps some statistics will make that point.*

- It took 50 years (1900–1950) for total annual U.S. energy consumption to go from 4 million barrels of oil equivalent per day to 16 million. It took only 20 years (1950–1970) to go from 16 to 32. (That trend is continuing, but we have slowed the rate recently.)
- With only 5% of the world's population, the United States consumes about $33\frac{1}{3}$% of its energy and also produces about $33\frac{1}{3}$% of the world's gross national product (GNP). The fact remains, though, that some nations such as Japan, West Germany, and Sweden produce the same or more GNP per capita with significantly less energy than the United States.
- Domestic oil production peaked in 1970, while domestic gas production peaked in 1974. Deregulation has improved our domestic production in the short run, but in the long run we will face continued decreasing domestic output.
- Prices per barrel for imported crude have rapidly escalated from $3.00 in the early 1970s to $12 in 1973–1974 and $32.00 in 1980.
- The United States has been a net importer of oil since 1947. In 1970 the bill for this importation was only $3 billion; by 1978 it was $42 billion; by 1979, $60 billion; and by 1980, $80 billion, even though the volume imported was less than in 1979. This has severely damaged our trade balance and weakened the dollar in national markets.
- Oil production in the Soviet Union (still not a net importer) is expected to peak in the early 1980s and the North Sea's fields will decline in output by 1984. Also, many emerging countries are dramatically increasing their consumption and are making demands for their share.

There are no easy answers, as the following demonstrate:

*The statistics here come from numerous sources, mostly governmental publications. Some also come from R. Stobaugh and D. Yergin, *Energy Future* (see the Bibliography).

- Many look to coal as the answer. Yet coal contains sulfur whose emissions damage human lungs and form acid rain that defaces buildings and kills life in lakes. Apparently this problem is just beginning. Coal, like other fossil fuels, develops large amounts of CO_2. Carbon dioxide tends to trap heat on the earth's surface, creating a greenhouse effect. The total effects are unknown but could involve substantial warming of the polar regions and creation of deserts in much of our country.

- Synfuels require strip mining, large costs, and demands for water in areas where there is little.

- Solar-generated energy (whether electricity through photovoltaics or thermal) is very expensive and has interesting legal problems; e.g., who owns the solar rights?

- Biomass energy is also expensive, and any sort of monoculture would require large amounts of land. Some fear total devastation of forests. At best, biomass can provide only a few percentage points of our total needs without large problems.

- Wind energy has technological "noise" and aesthetic problems that probably can be overcome, but again it is very expensive.

- Alcohol production from agricultural products raises perplexing questions about the use of food products for energy when large parts of the world are starving.

- Fission has the well-known problems of waste disposal, safety, and a short time span with existing technology. Without breeder reactors we will soon run out of fuel, and of course breeder reactors dramatically increase the production of plutonium—a nuclear bomb raw material.

- Fusion seems to be everyone's panacea or hope for the future, but many claim that we don't know the area well enough yet to predict its problems. When available commercially, fusion may very well have its own style of environmental-economical problems.

The preceding discussion paints a rather bleak picture. We don't mean to be alarmists, but our nation and the world are facing severe energy problems. There appears to be no simple answer.

Time and again energy management has shown that it can reduce energy costs (and consumption) subtantially. This freed energy can be used elsewhere so one energy source not already mentioned is energy management. In fact, energy available from energy management activities has proven to be the most economical source of "new" energy in most situations. Now, energy management can't solve the nation's problems, but *perhaps it can buy enough time and ease the strain on our environment so that we can develop new sources.*

1.1.3 Emotional

We all want to leave behind a world that is a little better than what we found when we arrived. Energy management allows us to feel like we have at least tried. Usually, energy management activities are more gentle to the environment than large-scale energy production and certainly lead to less consumption of scarce and valuable resources.

We feel energy management's time has come. The world will see a dire need for adequately trained engineers in the field of energy management. In fact, there are a large number of jobs available today. This text will help you prepare for your future career, which will be exciting and definitely challenging.

1.2 DESIGNING AN ENERGY MANAGEMENT PROGRAM

1.2.1 Management Commitment

The most important single ingredient for the successful implementation and operation of an energy management program is commitment to the program by top management. Without this commitment, the program will likely fail to reach its objectives.

Actually, there are two situations that are likely to occur with equal probability. In the first, management has decided that energy management is necessary and wants a program implemented. This puts you in the *response* mode. In the second, you might decide to convince management of the need for the program so you are in the *aggressive* mode. Obviously, the most desirable is the response mode as much work has been circumvented, but a large number of programs have been started through the aggressive mode. Let's consider these one at a time.

In a typical scenario of the response mode, management has seen rapidly rising energy prices and/or curtailments and has heard of the results of other energy management programs and then has initiated action to start the program. In this case, commitment already exists, and all that needs to be done is to cultivate that commitment periodically and to be sure the commitment is evident to all affected by the program. We will discuss this more when demonstrating the commitment.

In the aggressive mode, an engineer may have noted that energy costs are rising dramatically and that sources are less secure. Also, that engineer may have taken a course in energy management, attended professional conferences on the subject, and/or read papers. At any rate, the engineer is now convinced there is need for a program and all that remains is to convince management and obtain their commitment.

Management is convinced best through facts and statistics. Sometimes the most startling way to show the facts is through graphs such as Figure 1-1. Note that different goals of energy cost reduction are shown. This graph can be done in total for

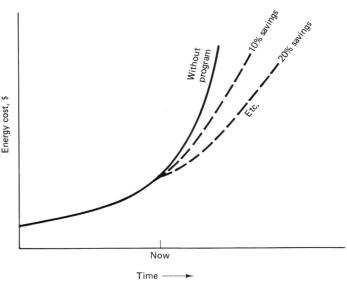

Figure 1-1 Energy costs—past and future.

all energy sources, or several graphs can be used, one for each source. The latter is probably better as savings goals can be identified as to energy source. It is important to have accurate data. Past figures can use actual utility bills, but future figures call for forecasting. Local utilities can help here, as can the various departments of energy at state levels.

Be sure to follow this quickly with quotes on programs from other companies showing these goals are realistic. Other company experiences are widely published in the literature, or results can be obtained through direct contacts with the energy manager in each company. Typical cost avoidance figures are shown in Table 1-1, but be careful. As time progresses and the technology matures, these figures tend to move up. For example, only a short time ago few people believed that an office building could reduce energy consumption by 70% or that manufacturing plants could operate on half the energy previously required, yet both have occurred recently.

The argument for an energy management program could then talk about the likelihood of energy curtailments or brownouts and what they could mean to the company. This could be followed by a discussion of what the energy management program could do to minimize the impacts of curtailments and brownouts.

Finally, the presentation should discuss the competition and what they are doing. Very good statistics on this can be obtained from trade and professional

TABLE 1-1 TYPICAL ENERGY SAVINGS OVER WHAT WOULD HAVE BEEN

Low cost, no cost changes	5–10%
Dedicated programs (3 years or so)	25–35%
Long-range goal	40–50%

organizations as well as the U.S. Department of Energy. The savings obtained by the competition can also be used in developing goals in Figure 1–1.

1.2.2 Energy Management Coordinator

For the vitality of the program, it is important to have one person who is responsible for its success or failure. Without this single person to turn to, management is likely to find that the energy management program has disappeared in the forest of other problems.

That person should be strong, dynamic, goals oriented, and a good manager. Most importantly, management should support that person with resources (including a staff—to be discussed).

Organizationally, the energy management coordinator (EMC) should report as high as possible without losing line orientation. In a multiplant or even multidivisional corporation, there will be several such coordinators—one for each plant and one for each level of organization. Typical scenarios are illustrated in Figure 1–2.

1.2.3 Backup Talent

Unfortunately, not all the talent necessary for a successful program resides in one person or discipline. For example, several engineering disciplines may be necessary to accomplish a full-scale study of the plant steam production, distribution, usage, and condensate return. For this reason, most successful energy management programs have provisions for one or more energy management committees. Actually, two types of committees may be necessary: technical and steering.

The technical committee is usually composed of several persons with good technical talent in their discipline. Chemical, industrial, electrical, civil, and mechanical disciplines as well as others may all be represented on this committee. Their responsibility is to provide backup for the coordinator and plant-level people when technical demands are too severe. For example, the committee will keep up with developing technology and research into potential applications company-wide. The results can then be filtered down.

While the energy management coordinator most likely works full time, the technical committee may be part time, being called upon as necessary. In a multiplant or multidivisional organization, the technical committee may also be full time.

The steering committee (actually the two committees can be combined into one energy management committee) has an entirely different purpose. It is to help guide the activities of the energy management program and to aid in communications through all organizational levels. The committee members are usually chosen so that all major areas are represented. A typical organization is presented in Figure 1–3.

Although this organization may be viable for one company, more than likely another company will have something different. Membership is chosen because of widespread interest and a sincere desire to aid in solving the energy problems. Often departmental and hourly representatives are chosen on a rotating basis.

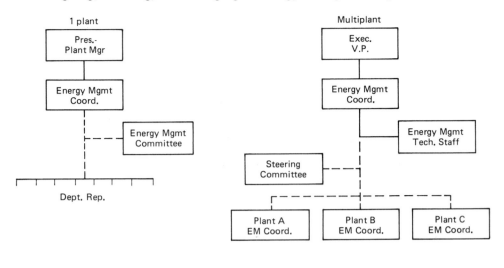

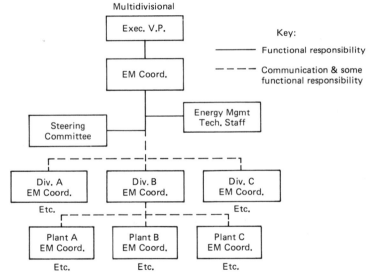

Figure 1-2 Typical organization designs for energy management programs.

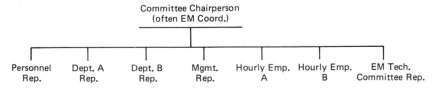

Figure 1-3 Energy management steering committee.

By the very nature of its makeup, the steering committee aids in speeding communications to all organization levels and ensures program sensitivity for plant personnel. Also, the committee should be able to develop a good composite picture of plant energy consumption which would help the energy management coordinator (EMC) choose and manage his/her activities.

1.2.4 Reporting and Monitoring

It's critical for the EMC and the steering committee to have their fingers on the "pulse of energy consumption" in the plant. This can best be done through an effective and efficient system of energy reporting. Later, we will discuss such a system of Btu accounting in some depth, but a brief introduction will be helpful here.

The objective is to measure the energy consumption and compare it to goals or some type of *standard energy consumption*. Ideally, this should be done for each operation cost center in the plant, but you are likely to find that the required metering devices simply do not exist. Indeed, you will be fortunate if any major energy consumers such as large motors, steam generators, etc., are submetered. Many plants only meter energy at one place—where the various sources enter the plant. Most plants are attempting to remedy this, however, by installing metering devices when the opportunity exists (steam system shutdowns, vacation downtimes, etc.). Items that need to be metered include steam, compressed air, and chilled and hot water.

As always, the reporting scheme needs to be reviewed periodically to ensure that no unneeded material is being generated, that all necessary data are available, and that the system is efficient and effective.

1.2.5 Training

Most energy management coordinators find that substantial training is necessary. This training can be broken down as shown in Figure 1–4.

Obviously, the training cannot be accomplished overnight, nor is it ever "com-

Personnel involved	Type of necessary training	Source of required training
1. Technical committee	1. Sensitivity to EM	1. In house (with outside help ?)
	2. Technology developments	2. Professional societies, universities, consulting groups, journals
2. Steering committee	1. Sensitivity to EM	1. See 1 above
	2. Other industries' experience	2. Trade journals, energy sharing groups, consultants
3. Plant-wide	1. Sensitivity to EM	1. In house
	2. What's expected, goals to be obtained, etc.	2. In house

Figure 1–4 Energy management training.

pleted.'' Staff changes occur, new technology develops, employees at all levels change, and production methods change. All these precipitate training or retraining. The energy management coordinator must assume responsbility for this training.

1.3 STARTING AN ENERGY MANAGEMENT PROGRAM

There are several items that contribute to the successful start of an energy management program. They include the following:

1. Visibility of start-up
2. Demonstration of management commitment
3. Good early project selection

1.3.1 Visibility of Start-up

As in most endeavors, an energy management program must have the backing of the people involved to be successful. Obtaining this support is often not an easy task, so careful planning is necessary. The people must

1. Understand why the program exists and what its goals are
2. See how it will affect their jobs and income
3. Know that the program has full management support
4. Know what is expected of them

Seeing that this information is transferred is a task of management and the energy management coordinator, and the transfer must be designed uniquely, with the company taking advantage of existing communications channels. All channels used must take into consideration the preceding four items. Some methods have proven to be useful in most companies:

- *Memos.* Memos announcing the program can be sent to all employees. A fairly comprehensive memo can be sent to all management personnel from first-line supervision up. This memo should give fairly complete details of the program. A more succinct one can be sent to all other employees that briefly states why the program is being formed and what is expected of them. These memos should be signed by local top management.
- *News releases.* Considerable publicity will often accompany the program start-up. Radio, TV, posters, newspapers, and billboards can all be used. The objective here is to obtain as much visibility for the program as possible and to reap any favorable public relations that might be available. News releases should, of course, contain information of interest to the general public as well as employees.

- *Meetings.* Corporate, plant, and department meetings are sometimes used, in conjunction with or in lieu of memos, to announce the program and provide some details. Top management can demonstrate commitment by being at these meetings. Time should be provided for interaction.

- *Films, video tapes.* Whether done in house or purchased, films and video tapes can be used to add another dimension to the presentation. They can also be reused for new employee training later.

1.3.2 Demonstration of Management Commitment

As mentioned earlier, management commitment is absolutely necessary, but this commitment must be obvious to all employees for the program to reach full potential. Management participation in the program start-up demonstrates this commitment, but some other things can be done to emphasize it:

- *Rewarding participating individuals.* Recognition is still extremely motivating to most employees. If an employee has been a staunch supporter of the program, this should be recognized by a "pat on the back," a letter in the files, acknowledgment at performance appraisal time, etc. This is especially true if the employee has made a suggestion that led to large energy savings. Such activities can be recognized through monetary rewards, publicity, or both. Publicity can be in the newsletter (see below), on bulletin boards, or in plant-department meetings.

- *Reinforcing commitment.* Periodically, management should reinforce its commitment, probably on a visibility scale lower than before. Here, existing newsletters or a separate one for the energy management program can be used by having a short column or letter from management on the results of the program to date and plans for the future. As mentioned, this same newsletter can be used to recognize outstanding suggestions from employees. Management must also realize that they are continually watched by employees. Saying commitment is not enough—it must be demonstrated.

- *Funding cost-effective proposals.* All companies have capital budgeting problems, in varying degrees of severity, and unfortunately energy projects are not front-line items such as equipment acquisition. However, management must realize that their energy management team is a dedicated group of professionals who usually are highly motivated. To refuse their proposals while accepting another with less economic attractiveness is a sure way to kill enthusiasm. Energy management projects need to compete with others fairly. If it is cost effective, it should be funded. If no money is available for capital expenditures, then management should make this obvious at the outset.

1.3.3 Early Project Selection

The energy management program is on treacherous footing in the beginning. Most employees wonder if their heating is going to be set back, their air conditioning turned off, and their lighting reduced. If one or more of these do occur, it's little wonder employee support wanes. These things *might* occur eventually,* but wouldn't it be smarter to have as early projects less controversial actions? Also, an early failure could be harmful if not disastrous to the program. Consequently, the astute energy management coordinator will "stack the deck" in his or her first set of projects. These projects should be ones with rapid payback, a high probability of success, and few negative consequences.

These end-of-the-rainbow projects aren't as rare and difficult to find as you might expect. Every plant has a few, and the energy management coordinator should be looking for them.

One example we are familiar with involved a rather dimly lit refrigerator warehouse area. Mercury vapor was the light source. The local energy management coordinator did a relamping project. He switched from mercury vapor to high pressure sodium (a significantly more efficient source) and carefully designed the system to improve the lighting levels. Savings were quite large (less energy for lighting, less "heat of light" that had to be refrigerated), and, most importantly, the employees liked it. Their environment was improved since light levels were higher than before.

Other examples include the following:

1. Repairing of steam leaks. (Even a small leak can be very expensive over a year and quite uncomfortable for employees working in the area.)
2. Insulating of steam, hot water, and other heated fluid lines and tanks. (Loss through an uninsulated steam line can be quite large, and the ambient air may be heated unnecessarily.)
3. Power factor improvement. (We will discuss this later, but in electrical energy management, something called the power factor can be improved by several in-house actions. This saves dramatically on the electrical utility cost and can improve voltage service throughout the plant—sometimes noticeably. No negative employee consequences occur, but the employees do have to be told what has happened since power factor improvement is not visible as in the lighting example.)

This list is, of course, incomplete, and what may be glamourous for one plant might not be for another. All plants, however, do have such opportunities. Of course, highly successful projects should be accompanied by publicity at all stages of the program especially at the beginning.

---

*As we said in the beginning, our goal is increased profits through management of energy. Any action that saves energy at the expense of widespread productivity is not likely to maximize profits.

1.4 MANAGEMENT OF THE PROGRAM

Creativity is a vital element in the successful execution of an energy management program, and management should do all it can to encourage creativity rather than stifle it. Normally, this implies a laissez-faire approach to management (with adequate monitoring). Management by objectives (MBO) is often utilized.

Goals need to be set, and these goals should be tough but achievable, measurable, and specific. Once management and the energy management coordinator have agreed on the goals, the coordinator should be left alone to do his or her job. Obviously, such an approach requires good monitoring or reporting systems, as will be discussed in the next section.

Examples of such goals might be the following:

- Total energy per unit of production will drop by 10% the first year and an additional 5% the second.
- Within 2 years all energy consumers of 5 million British thermal units per hour (Btuh) or larger will be separately metered for monitoring purposes.
- Each plant in the division will have an active energy management program by the end of the first year.
- All plants will have contingency plans for gas curtailments of varying duration by the end of the first year.
- All boilers of 50,000 lb/h or larger will be examined for waste heat recovery potential the first year.

The energy management coordinator must realize that the agreed upon goals are made to be met or exceeded and operate accordingly. This means quickly establishing reporting systems to measure progress toward the goals and developing strategy plans to ensure progress. Gantt or CPM charting is often used to aid in the planning and assignment of responsibilities.

Some concepts or principles that may aid the EMC in this execution are the following:

- Energy costs are to be controlled, not just Btus. This means that any action that reduces energy costs is fair game. Demand shedding or leveling is an example activity that saves dollars but not Btus (directly).
- Energy needs are to be recognized and billed as a direct cost, not as overhead. Until the energy flow can be measured and charged to operating cost centers, the program will not reach its ultimate potential.
- Only the main energy functions need to be metered and monitored. The Pareto or ABC principle states that the majority of the energy costs are incurred by only a few machines. They are the ones that should be watched carefully.

1.5 ENERGY ACCOUNTING

Energy accounting is keeping track of energy consumption and costs. It may range from a simple index of Btu/ft² or Btu/unit of production all the way to a complex standard cost system complete with variance reports. In all cases, energy accounting requires metering. To be able to monitor the energy flow through a cost center, no matter how large or small, requires the ability to measure incoming and outgoing energy. The lack of necessary meters is probably the largest single deterrent to the widespread utilization of energy accounting systems.

Energy User News * once printed an article with the headline "Accounting of Energy Seen Corp. Must." The same article went on to say the following:

> Successful corporate-level energy managers usually rank energy accounting systems right behind commitment from top corporate officials when they list the fundamentals of an ongoing energy conservation program. If commitment from the top is motherhood, careful accounting is apple pie.

All top energy management experts do seem to agree that energy accounting is vitally necessary for a vigorous energy mangement program. There, however, agreement seems to end. As in financial accounting, the level of sophistication or detail varies considerably between firms. Before describing several systems, let's examine these levels of sophistication.

1.5.1 Levels of Energy Accounting

A very close correlation can be developed between financial accounting levels of sophistication and those of energy. This is outlined in Figure 1–5.

Most companies with successful energy management programs have passed

Financial	Energy
1. General accounting	1. Effective metering, development of reports, calculation of energy efficiency indices
2. Cost accounting	2. Calculation of energy flows and efficiency of utilization for various cost centers; requires substantial metering
3. Standard cost accounting historical standards	3. Effective cost center metering of energy and comparison to historical data; complete with variance reports and calculation of reasons for variation
4. Standard cost accounting engineered standards	4. Same as 3 except now standards for energy consumption determined through accurate engineering models

Figure 1–5 Comparison between financial and energy accounting.

* *Energy User News,* Aug. 27, 1979, p. 1.

level 1 and are working toward the necessary submetering and reports systems for level 2. In most cases, the subsequent data are compared to previous years or to a particular base year. Few companies have, however, developed systems that will calculate variations and find causes for those variations (level 3). Two notable exceptions are General Motors and Carborundum. Their programs are described in some detail below. To our knowledge, no one has yet completely developed the data and procedures necessary for level 4, a standard Btu accounting system.

1.5.2 One-Shot Productivity Measures

The purpose of one-shot productivity measures is illustrated in Figure 1–6. Here the energy utilization index is plotted over time, and trends can be noted.

Significant deviations from the last period (or, more accurately, the same period as last year) can be noted and explanations sought. This is often done to justify energy management activities or at least to show their effect. For example, in Figure 1–6 an energy management (EM) program was started at the beginning of year 2. Its effect can be noted by comparing peak winter consumption in year 2 to that of year 1. This has been a good program (or a mild winter, or both).

Table 1-2 shows some indices often used. Advantages and disadvantages are noted, but the discussion is incomplete (the jury is still out) so some independent study is required for specific applications. Also, companies may be using some indices of which we're unaware.

Table 1-3 proposes some newer concepts. Advantages and disadvantages are shown, but since most of them have not been utilized in a large number of companies,

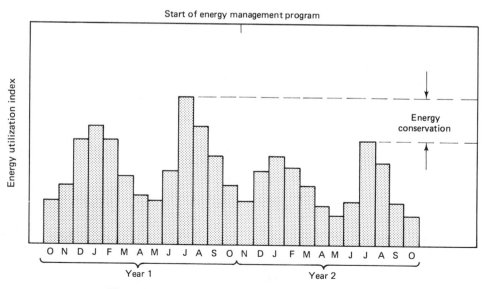

Figure 1–6   One-shot energy productivity measurement.

TABLE 1–2 COMMONLY USED INDICES

Productivity indicator	Advantages	Disadvantages
1. Btu/unit of production	1. Concise, neat 2. Often accurate when process needs high 3. Good for interplant and -company comparison when appropriate	1. Difficult to define and measure "units" 2. Often not acurate (high HVAC* and lighting makes energy nonlinear to production)
2. Btu/degree day	1. Concise, neat, best used when HVAC* is a majority of energy bill 2. Often accurate when process needs are low or constant 3. Very consistent between plants, companies, etc. (all mfg can measure degree days)	1. Often not accurate (disregards process needs) 2. Thermally heavy buildings such as mfg plants usually don't respond to degree days
3. Btu/ft^2	1. Concise, neat 2. Accurate when process needs low or constant and weather consistent 3. Very consistent (all mfg can measure square feet) 4. Expansions can be incorporated directly	1. No measure of production or weather 2. Energy not usually linearly proportional to floor space (piecewise linear?)
4. Combination, e.g., Btu/unit-degree day-ft^2 or Btu/unit-degree day	1. Measures several variables 2. Somewhat consistent, more accurate than above measures 3. More tailor-made for specific needs	1. Harder to comprehend

*Heating, ventilating, and air conditioning.

there are probably more of each. Also, there are an infinite number of possible indices, and only three are shown here.

1.5.3 General Motors: Energy Responsibility Accounting

General Motors states that a good energy accounting system is implemented in phases: (1) design and installation of accurate metering, (2) development of an energy budget, and (3) publication of regular performance reports including variances. These phases are examined in the following.

Phase 1: Metering.   For execution of a successful energy accounting program, it is absolutely necessary to be able to measure energy flow by cost center. The designing of cost center boundaries is a critical step and should be done carefully. There is a minimum and a maximum size for effectiveness, but the primary criterion is how much energy is involved. For example, a bank of large electric induction heat-treating furnaces might need separate metering even if the area involved is relatively small, but a large assembly area with few energy-consuming devices may not need separate

TABLE 1–3   PROPOSED INDICES

Productivity indicator	Advantages	Disadvantages
1. Btu/sales dollar	1. Easy to compute	1. Impact of inflation
2. $\dfrac{\$\ \text{energy}}{(\$\ \text{sales})\ \text{or}\ (\$\ \text{profit})\ \text{or}\ (\$\ \text{value added})}$	1. *Really what's desired* 2. Inflation cancels *or* shows changing relative energy costs 3. Shows energy *management* results, not just conservation (e.g., fuel switching, demand leveling, contingency planning)	1. Very complex, e.g., lots of variables affect profit including accounting procedures 2. Not good for general employee distribution
3. Btu/DL hour (or machine hour or shift) where DL = direct labor	1. Almost a measure of production (same advantage as in Table 1–2) 2. Data easily obtained when already available 3. Comparable between plants or industries 4. Good for high process energy needs	1. More complex, e.g., can't touch a DL hour like a unit of production 2. Energy often not proportional to labor or machine input, e.g., high HVAC and lighting

metering. Flexibility is vitally important since what is too small today may not be with tomorrow's energy costs. General Motors says that, in general, metering of areas smaller than 250,000 ft², may not be justified unless there are large energy consumers involved. Of course, the choice of meters used is important. Meters should be accurate, rugged, and cost effective, and have a good turndown ratio (the ability to measure accurately over the entire range of flow involved).

Having the meters is not enough. A system must be designed to gather the data, record them, and finally turn them into something useful. Meters can be read manually, they can record information on charts for permanent records, and/or they can be interfaced with microcomputers for real-time reporting and control. Too many systems fail because the data collection system is not being adequately utilized.

Phase 2: Energy Budget. The unique and perhaps vital part of General Motors' approach is the development of an energy budget. (By doing this, the energy responsibility accounting (ERA) is somewhere between levels 3 and 4 of Figure 1–5. If the budget is determined through engineering models, then it is a standard cost system and is at level 4.)

Basically, there are two ways to develop the energy budget: by statistical manipulation of historical data and by utilization of engineering models.

By using historical data, the statistical model shows how much energy was utilized and how it compared to the standard year(s), but it does not show how efficiently it was used.

For example, consider the data shown in Table 1–4.

In the statistical model, an assumption is made that the base years are charac-

TABLE 1–4 ENERGY DATA
FOR STATISTICAL MODEL[a]

	1980	1981	1982
Total energy (units)	1,000	1,100	1,050
Production (units)	600	650	650
Square feet	150,000	150,000	170,000
Degree days (heating)	6,750	6,800	6,800

[a] Taken, in part, from R. P. Greene, (see the Bibliography).

teristic for all future years. Consequently, if 1981 produced 600 units with the same square footage and degree days as 1980, 1000 units of energy would be required. If 970 were used, the difference (30 units) would be due to conservation.

We could use multiple linear regression to develop the parameters for our model, given as follows:

$$\text{energy forecast} = a(\text{production level}) + b(\text{ft}^2) + c(\text{degree days}) \qquad (1\text{--}1)$$

$$X_4 = aX_1 + bX_2 + cX_3$$

where X_1 = production (units) X_3 = weather data (degree days)
X_2 = floor space (ft^2) X_4 = energy forecast (Btu)

Of course, the actual factors included in the model will vary between companies and need to be examined carefully.

Multiple linear regression estimates a universe regression from a set of sample data. Using the base years, the procedure estimates values for parameters a, b, and c in Equation 1–1 so as to minimize the squared error where

$$\text{squared error} = \sum_{\substack{\text{base} \\ \text{years}}} (X_4^1 - X_4)^2 \qquad (1\text{--}2)$$

where X_4 = energy forecast by model
X_4^1 = actual energy usage

The development and execution of this model is beyond the scope of this book. For further development including an example, the reader is referred to any good statistics book.

Whatever method is used, the statistical model simply does not determine the amount of energy that should be used. It only forecasts consumption based on previous years' data.

The engineering model attempts to remedy this by developing complete energy balance calculations to determine the amount of energy theoretically required. By using the first law of thermodynamics, energy and mass balances can be completed for any process. The result would be the energy required for production. Similarly, HVAC and lighting energy needs could be developed using heat loss equations and other simple calculations. Advantages of the engineering model include improved accuracy and flexibility in reacting to changes in building structures, production schedules, etc. Also, computer programs exist that will calculate the needs for HVAC and lighting.

Phase 3: Performance Reports.   The next step is the publication of energy performance reports that compare actual energy consumption with that predicted by the models. The manager of each cost center can be evaluated on his or her performance as shown in these reports. The publication of these reports is the final step in the effort to remove energy cost from an overhead category to a direct cost or at least a *direct* overhead item. One example report is shown in Figure 1–7.

	Actual	Budget	Variance	% variance
Department A				
Electricity	2000	1500	+500	+33.3%
Natural gas	3000	3300	−300	−9.1%
Steam	3500	3750	−250	−6.7%
Total	8500	8550	−50	−0.6%
Department B				
Electricity	1500	1600	−100	−6.2%
Natural gas	2000	2400	−400	−16.7%
Fuel oil	1100	1300	−200	−15.4%
Coal	3500	3900	−400	−10.2%
Total	8100	9200	−1100	−11.9%
Department C				
.				
.				
.				

Figure 1–7   Energy performance report (10^6 BTU).

Sometimes more detail on variance is needed. For example, if consumption were shown in dollars, the variation could be shown in dollars also and broken into price and consumption variation. Price variation would be calculated as the difference between the budget and actual unit price times the present actual consumption. The remaining variation would be due to change in consumption and would be equal to the change in consumption times the budget price. This is illustrated in Figure 1–8. Other categories in variation could include fuel switching, pollution control, and new equipment. (This will be examined more in the next section.)

The example department (heat treating) shown in Figure 1–8 also portrays a common problem in energy management. The energy management program has been quite successful as the consumption (at old prices) was reduced by

$$\$500 + \$3980 + \$1151 = \$5631$$

Total energy cost, however, went up by

$$\$500 - \$1000 + \$1000 = \$500$$

	[A] Actual $	[B] Budget $	[C] Unit price (budget)	[D] Unit price (actual)	[E] $A - B^a$ variance	[F] $(D - C)A^b$ price variance	[G] $E - F$ or $(B^b - A^b)C$ consumption variance
Department (source)	10^6 Btu	10^6 Btu	$/10^6$ Btu	$/10^6$ Btu			
Heat treating (electricity)	$9,000 2,000	$8,500 2,125	$4.00 —	$4.50 —	+$500 —	+$1000 —	−$500 —
(natural gas)	15,000 4,808	16,000 6,400	2.50 —	3.12 —	−1000 —	+2980 —	−3980 —
(steam)	10,000 2,242	9,000 2,571	3.50 —	4.46 —	+1000 —	+2151 —	−1151 —
(total)	$34,000	$33,500	—	—	+$500	+$6131	−$5631

^{a} Measured in $

^{b} Measured in 10^6 Btu

Figure 1-8 Energy cost in dollars by department with variance analysis.

This is due to a substantial price variation of

$$\$1000 + \$2980 + \$2151 = \$6131$$

Consequently, energy costs did go up but only to $34,000 instead of to

$$2151(4.50) + 6400(3.12) + 2571(4.46) = \$40,997$$

had consumption not been reduced. Total cost avoidance then was

$$\$40,997 - \$34,000 = \$6997$$

which is the drop in consumption times the actual price or

$$(2125 - 2000)4.5 + (6400 - 4808)3.12 + (2571 - 2242)4.46 = \$6997$$

1.5.4 Carborundum Accounting System

The Carborundum system carries the analysis of variance several steps further by showing price variation as before, but consumption variance is broken down into volume-mix, weather, pollution, conservation, alternative fuels, and others. By stating the consumption variation in this manner, the Carborundum system provides more information for management action. The net impact would still be the algebraic sum of all effects.

It is important to note that no accounting system is a panacea and that any system is only as accurate as the metering and reporting systems allow it to be. Calculations for the different variances are actually quite simple. The calculation procedures are as follows.

Product Mix Effect.   For each product line, the effect of changing production volume on energy consumption for a given source is

$$PM_{ij} = (Q_i - Q_i^1)E_{ij}^1 \qquad (1\text{-}3)$$

where PM_{ij} = product mix effect (10^6 Btu) of changing production levels on product i for energy source j
Q_i = quantity of product i made to date
Q_i^1 = quantity of product i made base year to date (or standard quantity)
E_{ij}^1 = amount of energy of type j (10^6 Btu) required per unit of product i base year (or standard requirement)

To specify this in dollars, simply multiply PM_{ij} by last year's cost of energy of type j:

$$PM_{ij}(\$) = PM_{ij}C_j^1 \qquad (1\text{-}4)$$

where C_j^1 = cost of energy type j in base year (or standard price)

Heating Effect.   The effect of varying heating demands can be determined from the following equation:

$$H_j = (DD_H - DD_H^1)E_{Hj}^1 \qquad (1\text{-}5)$$

where H_j = heating effect on consumption of energy type j (10^6 Btu) this year
DD_H = heating degree days this year to date
DD_H^1 = heating degree days base year to date (or standard quantity)
E_{Hj}^1 = quantity type j energy required per degree day base year to date (or standard quantity)

To specify this in dollars, simply use

$$H_j(\$) = H_j C_j^1 \qquad (1\text{-}6)$$

Cooling Effect.   The effect of varying cooling needs can be determined from

$$K_j = (DD_c - DD_c^1)E_{cj}^1 \qquad (1\text{-}7)$$

where K_j = cooling effect on consumption of energy type j (10^6 Btu) this year
DD_c = cooling degree days this year to date
DD_c^1 = cooling degree days base year to date (or standard quantity)
E_{cj}^1 = quantity of type j energy required per cooling degree day base year to date (or standard quantity)

This can be stated in dollars through

$$K_j(\$) = K_j C_j^1 \qquad (1\text{-}8)$$

Other Effects.   The effect of any other changes can be determined from the following equations repeated for as many effects as desired (e.g., pollution control or one general account "other"):

$$OE_j = O_j - O_j^1 \qquad (1\text{-}9)$$

$$OE_j(\$) = O_jC_j^1 \qquad (1\text{-}10)$$

where OE_j = effect of other changes on energy type j
O_j = energy type j effects from other changes
O_j^1 = energy type j effects from other changes base year to date (or standard)

Example

Ace Manufacturing Company has an active energy management program underway and has recently installed an energy accounting system of the type just developed. For the data given below comparing this year to last, calculate all variances and show the net impact:

	This year	Last year (base)
Product line 1 ($i = 1$)	1000 units	900 units
Product line 2 ($i = 2$)	2000 units	2200 units
Heating degree days	3570 °F days	3800 °F days
Cooling degree days	2200 °F days	2000 °F days
Price/10^6 Btu (nat. gas)	$3.50	$3.00
Price/10^6 Btu (electricity)	$5.00	$4.75
Nat. gas/heating degree day ($j = 1$)	25 $\times 10^6$ Btu	30 $\times 10^6$ Btu
Nat. gas/cooling degree day	0	0
Elec./heating degree day ($j = 2$)	0	0
Elec./cooling degree day	60 $\times 10^6$ Btu	70 $\times 10^6$ Btu
Gas required/unit product 1 (E_{11})	80 $\times 10^6$ Btu	100 $\times 10^6$ Btu
Gas required/unit product 2 (E_{21})	45 $\times 10^6$ Btu	50 $\times 10^6$ Btu
Elec. required/unit product 1 (E_{12})	400 $\times 10^6$ Btu	500 $\times 10^6$ Btu
Elec. required/unit product 2 (E_{22})	200 $\times 10^6$ Btu	300 $\times 10^6$ Btu

Product mix effect:

$$\begin{aligned} PM_{11} &= (Q_1 - Q_1^1)E_{11}^1 \\ &= (1000 - 900)(100 \times 10^6) \\ &= +10{,}000 \times 10^6 \text{ Btu} \end{aligned}$$

$$\begin{aligned} PM_{12} &= (Q_1 - Q_1^1)E_{12}^1 \\ &= 100(500 \times 10^6) \\ &= +50{,}000 \times 10^6 \text{ Btu} \end{aligned}$$

$$\begin{aligned} PM_{21} &= (Q_2 - Q_2^1)E_{21}^1 \\ &= (2000 - 2200)(50 \times 10^6) \\ &= -10{,}000 \times 10^6 \text{ Btu} \end{aligned}$$

$$\begin{aligned} PM_{22} &= (Q_2 - Q_2^1)E_{22}^1 \\ &= -200(300 \times 10^6) \\ &= -60{,}000 \times 10^6 \text{ Btu} \end{aligned}$$

$$PM_{11}(\$) = PM_{11}C_i^1$$
$$= (10{,}000 \times 10^6 \text{ Btu}) (\$3.00/10^6 \text{ Btu})$$
$$= +\$30{,}000$$

$$PM_{12}(\$) = PM_{12}C_2^1$$
$$= (50{,}000 \times 10^6 \text{ Btu}) (\$4.75/10^6 \text{ Btu})$$
$$= +\$237{,}500$$

$$PM_{21}(\$) = PM_{21}C_1^1$$
$$= (-10{,}000 \times 10^6 \text{ Btu}) (\$3.00/10^6 \text{ Btu})$$
$$= -\$30{,}000$$

$$PM_{22}(\$) = PM_{22}C_2^1$$
$$= (-60{,}000 \times 10^6 \text{ Btu}) (\$4.75/10^6 \text{ Btu})$$
$$= -\$285{,}000$$

net product mix effect $= -\$285{,}000 - \$30{,}000 + \$237{,}500 + \$30{,}000 = -\$47{,}500$

Heating effect:

$$H_1 = (DD_H - DD_H^1)\, E_{H1}^1$$
$$= (3570 - 3800) (30 \times 10^6 \text{ Btu})$$
$$= -6900 \times 10^6 \text{ Btu}$$

$$H_2 = 0$$

$$H_1(\$) = H_1 C_1^1$$
$$= (-6900 \times 10^6 \text{ Btu}) (\$3.00/10^6 \text{ Btu})$$
$$= -\$20{,}700$$

Cooling effect:

$$K_1 = 0$$

$$K_2 = (DD_c - DD_c^1)\, E_c^1$$
$$= (2200 - 2000)(70 \times 10^6 \text{ Btu})$$
$$= 14{,}000 \times 10^6 \text{ Btu}$$

$$K_2(\$) = K_2 C_2^1$$
$$= (14{,}000 \times 10^6 \text{ Btu}) (\$4.75/10^6 \text{ Btu})$$
$$= \$66{,}500$$

Other consumption effects: None shown.

Price effects – natural gas: The price effect for natural gas is the difference in prices between this year and the base year times the quantity used this year (see the Section 1.5.3). The quantity of natural gas used this year is

(3570 degree days) $(25 \times 10^6$ Btu/degree day) $+$ (1000 units product 1) $\times (80 \times 10^6$ Btu/unit) $+$ (2000 units product 2) $(45 \times 10^6$ Btu/unit)

$$= 259{,}000 \times 10^6 \text{ Btu natural gas}$$

$$\text{natural gas price effect} = (259,250 \times 10^6 \text{ Btu})(\$3.50 - \$3.00)/10^6 \text{ Btu}$$
$$= +\$129,625$$

Electricity: Similarly the price effect for electricity is the consumption this year times the price differential. The consumption this year is

(2200 degree days)(60 × 10⁶ Btu/degree day) + (1000 units product 1)
(400 × 10⁶ Btu/unit) + (2000 units product 2)(200 × 10⁶ Btu/unit)
$$= 932,000 \times 10^6 \text{ Btu}$$

$$\text{electricity price effect} = (932,000 \times 10^6 \text{ Btu})(\$5.00 - 4.75)/10^6 \text{ Btu}$$
$$= \$233,000$$

$$\text{total price effect} = \$233,000 + \$129,625$$
$$= +\$362,625$$

The total energy cost for this year is

(259,250 × 10⁶ Btu nat. gas)($3.50/10⁶ Btu) + (932,000 × 10⁶ Btu elec.) ($5.00/10⁶ Btu)
$$= \$5,567,375$$

To determine the total energy cost for the base year, we need the total consumption:

total consumption natural gas = (3800 degree days) (30 × 10⁶ Btu/degree day) + (900 units product 1) (100 × 10⁶ Btu/unit) + (2200 units product 2)(50 × 10⁶ Btu/unit)
$$= 314,000 \times 10^6 \text{ Btu}$$

total consumption electricity = (2000 degree days) (70 × 10⁶ Btu/degree day) + (900 units product 1) (500 × 10⁶ Btu/unit) + (2200 units product 2)(300 × 10⁶ Btu/unit)
$$= 1,250,000 \times 10^6 \text{ Btu}$$

total energy cost (base year) = (314,000 × 10⁶ Btu gas)($3.00/10⁶ Btu gas) + (1,250,000 × 10⁶ Btu elec.) ($4.75/10⁶ Btu)
$$= \$6,879,500$$

A summary of effects is given in Table 1–5.

TABLE 1–5 ENERGY EFFECTS BREAKDOWN

	Base year	This year	Change
Energy consumption ($)	6,879,500	5,567,375	−$1,312,125
Price effect	—	—	+$362,625
Cooling effect	—	—	+$66,500
Heating effect	—	—	−$20,700
Product mix effect	—	—	−$47,500
Conservation (balancing amount)	—	—	−$1,673,050

As can be seen from Table 1-5, the price and cooling effect changes were largely negative and offset the savings from heating and the product mix. Yet the total energy cost *dropped* by $1,312,125. Therefore, the balancing amount is due to conservation ($1,673,050). This is a large saving and signifies a successful program.

1.6 SUMMARY

This chapter has examined the design, initiation, and management of energy management programs. There is a strong emphasis in the chapter on energy accounting, especially cost center accounting and necessary submetering.

First, we define an energy management activity as any decision that involves energy and affects the profit level. Anything that improves profits and/or enhances competitive positions is good energy management, and anything else is poor energy management. The motivation for starting energy management programs is multi-faceted and varies among companies. However, the following outline lists the major reasons:

- Economic
  1. Energy management will improve profits and enhance competitive positions.
- National good
  2. Energy management is good for the U.S. economy as the balance of payments becomes more favorable and the dollar stronger.
  3. Energy management makes us less vulnerable to energy cutoffs or curtailments due to political unrest or natural disasters elsewhere.
- Emotional
  4. Energy management is kind to our environment as it eases some of the strain on our natural resources and may leave a better world for future generations.

In designing an energy management program, some ingredients are vitally necessary:

- *Top management commitment.* Commitment from the top must be strong and highly visible.
- *One-person responsibility.* The responsibility for the energy management program must lie in one person reporting as high in the organization structure as possible.
- *Committee backup.* The energy management coordinator must have, as backing, two types of talent in one or two committees. The first is steering, where direction for the program is provided. The second is technical, where technical backup is provided in the necessary engineering disciplines.

- *Reporting and monitoring.* An effective monitoring and reporting system for energy consumption must be provided.
- *Training.* Energy management is a unique undertaking. Hence, training and retraining at all levels are required.

To start an energy management program successfully, some fanfare must accompany the early stages. This can come through news releases, films, plant meetings, or a combination of them. Then early project selection is a critical step. Early projects should be visible, have good returns, and have few negative consequences.

Management is always a critical component of any program, and energy management with the needed personnel creativity is no exception. Tough, specific, and measurable goals need to be established. Once established, management should carefully monitor and let the energy management staff perform its functions. Staff and management need to realize that (1) energy costs are to be controlled—not consumption, (2) energy should be a direct cost—not an overhead item, and (3) only the main energy consumers need be metered and monitored closely.

Energy accounting is the art and science of tracing Btu and energy dollar flow through an organization. Cost center orientation is important, as is comparison to some standard or base and calculations of variances. Causes for these variances must then be sought. General Motor's energy responsibility accounting system is discussed in some depth, as is Carborundum's rather sophisticated approach to determining variances.

BIBLIOGRAPHY

BARRETT, JOHN A., "Energy Accounting Systems for Existing Buildings," *ASHRAE Journal,* Jan. 1979, p. 56.

"Btu Accounting," client newsletter, Seidman and Seidman, certified public accountants, Grand Rapids, 1980.

DUNCAN, ACHESON J., *Quality Control and Industrial Statistics,* Richard D. Irwin, Inc., Homewood, Ill., 1959.

Energy Conservation and Program Guide for Industry and Commerce, GPO, Washington, D.C., Sept. 1974.

FEARING, WATSON B., 3M Corp., St. Paul, personal correspondence, 1980.

GREENE, RICHARD P., "Energy Responsibility Accounting, an Energy Conservation Tool for Industrial Facilities," unpublished working paper, General Motors Corp.

HOWELL, EDWARD D., "Energy Accounting and Analysis: The Carborundum System," unpublished working paper, the Carborundum Company.

SHIRLEY, JAMES M., and W. C. TURNER, "Industrial Energy Management—First Steps Are Easy," *IE,* May 1978, pp. 34–41.

STOBAUGH, ROBERT, and DANIEL YERGIN, *Energy Future,* Random House, New York, 1979.

TURNER, W. C., and B. W. TOMPKINS, "Productivity Measures for Energy Management," in *IIE National Meeting, Detroit, Mich., Proceedings, May 1981.*

QUESTIONS

1.1. What is energy management? State the role of conservation in an effective energy management program.

1.2. Why is there an increasing interest in energy management?

1.3. In the concept of energy management, distinguish between an energy management steering committee and an energy management technical committee. Should they be combined into one committee or not?

1.4. In your opinion, what is the single most important ingredient for a successful energy management program?

1.5. You have recently been hired as a consultant to develop an energy cost accounting system for a medium-sized job shop plant involved in metal working (level 2 system). Discuss your approach to this project (briefly). State some of your first activities.

1.6. Discuss why you think a good energy accounting system is or is not needed for an effective energy management program.

PROBLEMS

1.1. In Section 1.3.3 you found that early project selection can be critical. For your university, state some energy management projects that might be good "first ones."

1.2. Again for your university, assume you are starting a program and are defining goals. What are some potential first-year goals?

1.3. As a new energy manager, you have been asked to predict the energy consumption for electricity for next month (February). Assuming consumption is dependent on units produced, that 1000 units will be produced in February, and that the following data are representative, determine your estimate for February.

Last year	Units produced	Consumption (kwh)
January	600	600
February	1500	1200
March	1000	800
April	800	1000
May	2000	1100
June (vacation)	100	700
July	1300	1000
August	1700	1100
September	300	800
October	1400	900
November	1100	900
December (1-week shutdown)	200	650
January	1900	1200

1.4. For the same data as given in Problem 1.3, what is the fixed energy consumption (at zero production, how much energy is consumed and for what is that energy used)?

1.5. For the following data, calculate the variances (in Btu and dollars) due to product mix changes, heating degree days, cooling degree days, price, and conservation.

	This Year	Base year
Product A	2000 units	1500 units
Heating degree days	4000°F days	4300°F days
Cooling degree days	2000°F days	2000°F days
Price/10^6 Btu gas	$4.00	$3.25
Gas/heating degree day	15×10^6 Btu	20×10^6 Btu
Elec./heating degree day	0	0
Gas/cooling degree day	0	0
Elec./cooling degree day	40×10^6 Btu	50×10^6 Btu
Gas/unit product A	30×10^6 Btu	40×10^6 Btu
Elec./unit product A	20×10^6 Btu	25×10^6 Btu
Price/10^6 Btu Elec.	$6.00	$5.00

1.6.   At the ACE Company, switching (due to a shortage of fuel oil) caused an increase in gas consumption as follows:

	Expected	Actual (due to switching)
Gas/heating degree day	40×10^6 Btu	45×10^6 Btu
Gas/unit product A	23×10^6 Btu	24×10^6 Btu
Gas/unit product B	14×10^6 Btu	16×10^6 Btu

The base year cost of natural gas is $2.80/$10^6$ Btu, while this year's cost is $3.50/$10^6$ Btu. *Determine the cost of fuel switching in Btus and dollars.* Heating degree days and production data are as follows:

Heating degree days: 1000°F days

Production: Product A: 500 units

Product B: 800 units

(*Note:* The preceding data concern only those units and heating degree days where fuel switching actually occurred.)

Chapter 2
The Energy Audit

2.0 INTRODUCTION

Energy management can be divided into the three processes of analysis, action, and monitoring. Analysis is the recognition of a problem, for example, the realization that production capacity is being limited as a result of insufficiency of the energy supply. Action is the process whereby some solution is chosen for the problem and steps are taken to implement the solution. Monitoring is the comparison between planned and actual progress toward chosen objectives. Depending on the results of the monitoring step, another action may be initiated, and the analysis, action, and monitoring sequence may again be repeated.

The energy audit is a part of the action step—a detailed analysis of the energy aspects of a cost or production problem, together with proposed solutions and objectives for use in monitoring. An energy audit is the collection and analysis of data on present energy use, the choice of energy management objectives and of specific measures to meet those objectives, and the process used to monitor progress toward those objectives.

In performing an energy audit, a person first performs the *analysis* phase—a detailed examination of data to determine where energy is used and how the use varies with time. Each of the physical systems within the facility is then inspected carefully and the results noted for future use. In the next—or *action*—phase, a team is assembled and performs walk-through inspections, specific goals are agreed upon, and action is initiated. The third phase is *monitoring*. Monitoring yields information on the degree to which chosen goals have been accomplished and shows where measures have been successful or have failed. This phase thus leads back to the initiation of more analysis, action, and monitoring. These phases are discussed in detail in this chapter.

2.1 ANALYSIS

The analysis phase of an energy audit consists of, first, examining historical and descriptive data for the facility; second, performing a detailed inspection of the eight major systems that make up a facility; and, third, collecting ideas for remedying deficiencies that have been noted. The result of this phase is a complete description of the time-varying energy consumption patterns of the facility, a list of all equipment that affects the energy consumption together with an assessment of its condition, a chronology of usual operating and maintenance practices, and a list of possible energy management ideas for future investigation.

2.1.1 Analysis of Bills

The first step in the analysis procedure is a detailed analysis of past energy bills. This is important for several reasons: The amount spent on energy puts an automatic upper limit on the amount that can be saved, the bills give the relative importance of various energy sources in the total energy bill, and an examination of where energy is used can point out previously unknown energy wastes. These data can be conveniently entered onto a form such as shown in Figure 2–1. Note that the most significant billing factors are shown, including peak demand for electricity. These data are then analyzed by energy source and billed location and tabulated or graphed as in Figure 2–2.

Location _____

From _____ to _____

	Electrical use			Gas use			Fuel oil	
Month	Peak kw	Usage: kwh	Cost	MMCF[a]	Dth[b]	Cost	Gallons	Cost
___	___	___	___	___	___	___	___	___
___	___	___	___	___	___	___	___	___
___	___	___	___	___	___	___	___	___
___	___	___	___	___	___	___	___	___
___	___	___	___	___	___	___	___	___
___	___	___	___	___	___	___	___	___

[a] 1 MMCF = 10^6 ft^3

[b] 1 Dth = 10 therms = 10^6 Btu

Figure 2–1   Summary form for energy use.

Figure 2–2 shows a very useful kind of analysis. From this figure, it is evident that most of the gas is used by the main heating plant. It therefore makes good sense to concentrate most of the energy management effort and money on the main heating plant. Several other items of interest can also be gleaned from these data. First note that the greenhouse uses a lot of energy. Subsequent investigation revealed that the heating and cooling were controlled by different thermostats and that the cooling was turned on when the temperature got too high—before the heat was turned off! Another problem revealed was the amount of energy used by the president's house.

Energy use – natural gas

Heating plant	$38,742.34	83.2% (cost)
East dormitory	4,035.92	8.7%
Married student apts	1,370.79	2.9%
U.G. dormitory	768.42	1.7%
Greenhouse	560.21	1.2%
Child development ctr	551.05	1.2%
President's home	398.53	0.9%
Art barn	104.77	0.2%

Figure 2–2   Gas use by metered location.

Great care must also be taken that the actual sources of energy consumption are known rather than assumed. For this purpose, a brief inspection was made to assure that the gas consumption meter at the greenhouse was not measuring gas consumption elsewhere as well; if the gas usage from several buildings had been metered at the same meter as the greenhouse, it would have been necessary to determine some way to allocate gas consumption to each one.

Another graph that should be made for each type of energy is shown in Figure 2–3. Each area of the country and each different industry type has its unique pattern of energy consumption, and the data presented in Figure 2–3 help to define and analyze these patterns. In the place where this example came from, much electrical energy is used in the summer for air conditioning, so the August electrical peak is not surprising (note that a knowledge of the timing and magnitude of this peak can be important if the peak demand charge is based on the *annual* peak).

2.1.2 Detailed Preliminary Inspections: Walk-Through Audits

A walk-through audit should systematically examine the major systems within a facility, using portable instrumentation and common sense guided by an anticipation of what can go wrong. The eight major systems, described in detailed below, are the building envelope; the heating, ventilating, and air conditioning system; the electrical

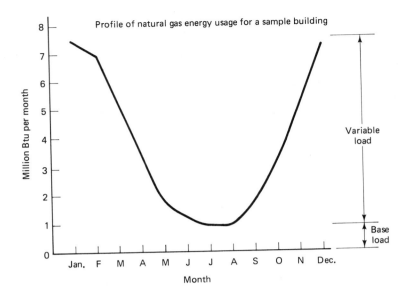

Figure 2-3   Graph of energy consumption over time. (Courtesy of U.S. Department of Energy Conservation, *Instructions for Energy Auditors,* DOE/CS-0041/12 and 0041/13, Sept. 1978.)

system; the lighting system, including all lights, windows, and adjacent surfaces; the hot water distribution system; the compressed air distribution system; and the manufacturing system. These systems together account for all the energy used in any facility; examining all of them is a necessary step toward understanding and managing energy utilization within a company.

Each of these systems can be examined with the help of the following questions.

1. What function(s) does this system serve?
2. How does this system serve its function(s)?
3. What are the indications that this system is working?
4. If this system is not working, how can it be restored to working condition?
5. How should this system be maintained?
6. Who has direct responsibility for the condition of this system?

As each system is inspected, forms similar to those of Figure 2-4 should be filled out. The following system descriptions can aid in the inspection process.

2.1.2.1 The building envelope.

The building envelope includes all building components that are directly exposed to the outside environment. Its main function is to protect employees and materials from outside weather conditions and temperature variations; in addition, it provides privacy for the business and can serve other psychological functions. The components of the building envelope are outside doors, windows, and walls; the roof; and, in some cases, the floor.

In inspecting the building envelope, look for drafts. Drafts can be caused by doors or windows that do not close completely, by broken windows, by holes in walls, and by cracks in walls. In one building examined by the authors, a new section was added onto an old building, and the two separated with time. Drafts are also a major source of personal discomfort, and removing them can create support for an energy management program. The cost of leaks of conditioned air can be calculated using the following formula:

$$
\begin{aligned}
\text{Annual cost} &= \$/\text{Btu} \times \text{ number of cubic feet per minute (cfm) through hole} \\
&\quad \times \text{ HDD (days} \times {}^\circ\text{F)} \times [\text{specific heat of dry air (Btu/1b} \cdot {}^\circ\text{F)} \\
&\quad + \text{ percent water vapor (lb moisture/lb dry air)} \\
&\quad \times \text{ specific heat of water vapor]} \times \text{ air density (lb/ft}^3) \\
&\quad \times \text{ 1440 min/day}
\end{aligned}
$$

$$
\begin{aligned}
&= \$/\text{Btu} \times \text{cfm} \times \text{HDD} \times (.24 + .013 \times .489) \\
&\quad \times .0766 \times 1440
\end{aligned}
\tag{2-1}
$$

System: Envelope

Component	Location	Maintenance condition	Est. air gap (total)
Door	North side	Poor	0.2 ft^2
	South	OK	0.05 ft^2
	Gymnasium	Good	None
Windows	North	Some broken	2.2 ft^2
	East	OK	None
Roof	Main building	No insulation	

(a)

Area	Type of lighting (e.g., HPS)	Watts per fixture	Number of fixtures	Total kw	Operating hours	Operating days	kwh/month
Interior							
Exterior							

(b)

Figure 2-4   (a) Completed inspection form for envelope. (b) Data collection form for lighting system.

where HDD = number of heating degree days per year
  .0766 = air density (at sea level—less at higher elevations)
  .24 = specific heat of dry air
  .013 = lb water vapor per lb dry air at 80°F and 60% relative humidity
  .489 = specific heat of water vapor

Also inspect the insulation on the outside walls and the roof. The amount of insulation in walls can often be determined by removing the plate from an electrical outlet (after ensuring that the fuse or circuit breaker for the outlet has been disconnected so that the outlet is no longer dangerous). The roof should also be inspected, both for insulation and for holes. The cost of heat loss through a wall or roof can be estimated from

$$\text{annual cost} = \$/\text{Btu} \times \text{area (ft}^2) \times \text{thermal conductance } c \text{ (Btu/h} \cdot \text{ft}^2 \cdot °\text{F)}$$
$$\times \text{ HDD (days} \times °\text{F)} \times 24 \text{ h/day} \quad (2\text{-}2)$$

where the thermal conductance c is the reciprocal of the R value for the roof or wall. (See Chapter 11 for more on insulation.)

Any doors from inside conditioned spaces to the outside are openings in the envelope, and keeping such doors closed is a worthwhile measure when energy is being used to heat or cool the interior. A good example of such an opportunity and its remedy are shown in Figure 2-5, a typical loading dock to which door seals have been added. Depending on the temperature difference between the outside and the inside, such dock seals can pay for themselves in 6 months or less.

2.1.2.2 The boiler and steam distribution system.

A boiler converts fuel into steam; the steam distribution system takes the steam from the boiler to the point of use. Boilers consume most of the fuel used in most production facilities. The boiler is thus the first place to look when attempting to reduce natural gas consumption. The steam distribution system is also a very important place to look when saving energy, since every pound of steam lost is a pound of steam that must be produced in a boiler. Boilers and the steam distribution system are covered in detail in Chapter 8. In this chapter, the emphasis is on determining the general condition of those systems in order to see whether more detailed investigation is warranted.

A boiler can explode, so it is of utmost importance that all the safety features be in operating condition. Most building safety codes require that a boiler be inspected periodically; make sure that these inspection requirements have been complied with. Also make sure that outside air can get to the boiler to provide oxygen for combustion. (Insufficient combustion air causes incomplete combustion and the generation of carbon monoxide. This is an inefficient use of energy and potentially dangerous.)

In an energy audit, there are some clear indications of opportunities to save energy and money. In a visual inspection of a boiler, look at the gauges first. If the gauges don't work, the boiler control system is probably not functioning correctly, and the opportunity exists for a substantial saving in fuel costs. If the most recent professional inspection and adjustment of the boiler occurred more than 2 years ago, the same opportunity is present. If the boiler stack temperature is more than 150°F

Figure 2-5   Loading dock seal. (Courtesy of Medalist Canvas Products.)

above the boiler water temperature, the boiler is operating less efficiently than necessary, and a professional adjustment is usually worthwhile. Also look for rust in water gauges and for boiler exhaust gas that is either black (too little outside air is being used in combustion) or clear (too much outside air).

The steam distribution system should be inspected when the boiler is inspected. The main components of the steam distribution system are steam lines, valves, and traps and condensate return lines, pumps, and tanks. Steam leaks can be expensive because of the amount of steam that is lost; condensate leaks represent lost heat and loss of treated water. In many environments, steam leaks can be detected by their hissing or by a directed cone of steam; in very noisy environments, it may be necessary to use an industrial stethoscope. Evidence of possible condensate leaks includes pools of hot water, dripping pipes, and rust spots on pipes. While looking for leaks, put your hand close to the pipes. If a pipe is too hot to touch, it probably should be insulated (see Chapter 11). The costs of steam leaks can be estimated by

$$\text{cost/year} = \$/\text{Btu} \times \text{lb steam lost/h} \times \text{Btu/lb} \times \text{operating h/year} \qquad (2\text{-}3)$$

where the number of pounds of steam lost per year can be either measured directly or estimated on the basis of experience, where the number of Btu/lb is the specific enthalpy of the steam and depends on the pressure of the steam as shown in Table 8-1, and where the number of operating hours per year comes from line management personnel.

Steam traps can provide a source of major cost savings. Two kinds of traps are shown in Figures 2-6 and 2-7. The main functions of a steam trap are to keep steam from escaping to the air, to drain condensate from steam lines, and to remove air and dissolved oxygen from condensate. These functions, described in more detail in Chapter 10, are usually accompanied by a characteristic clicking and, depending on the installation, by a visible discharge of a small amount of flash steam—condensate that evaporates upon contact with air. The clicking can be detected either by an industrial stethoscope or by putting one end of a long screwdriver on the steam trap and the other end next to an ear. (If the steam trap is carrying a small amount of steam relative to its capacity, the interval between clicks may be 5–20 min.) A sure sign that a steam trap is not working is the presence of live steam—steam that is under

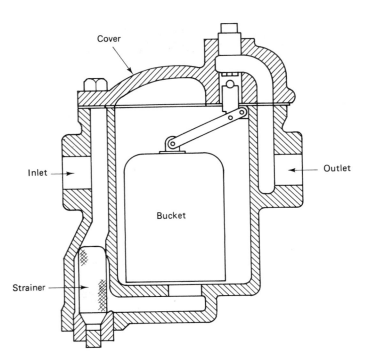

Figure 2-6 Inverted bucket steam trap. (Courtesy of Armstrong Machine Works.)

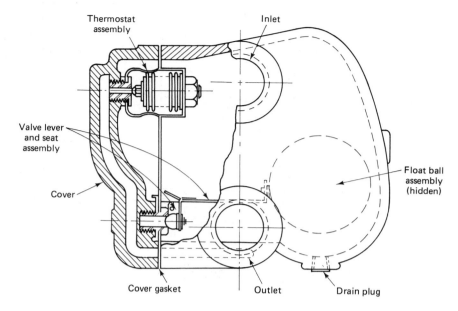

Figure 2-7   Float and thermostatic steam trap. (Courtesy of Armstrong Machine Works.)

pressure—blowing into the air, where a leak is not present. If escaping steam is live downstream from a trap, the trap has failed open and is letting steam through along with the condensate; if the escaping steam is live at the condensate return tank, one or possibly all of the steam traps have failed. (The calculation of costs of steam trap failures is the same as that for steam leaks.) Another problem caused by failed steam traps is water hammer—the phenomenon caused when steam pushes a wave of condensate against a bend in the steam distribution system, generally with a loud noise. Water hammer occurs when condensate is not being removed properly by steam traps. (Occasionally the system was not designed correctly. This fault should not be suspected unless the steam traps have been checked and found to be working.)

In addition to looking for leaks and checking steam traps, whoever is inspecting this system should check condensate return tanks to see whether insulation is required and should examine the pumps for leaks and to see whether the bearings are excessively noisy.

2.1.2.3 The heating, ventilating, and air conditioning system.   This system, an example of which is given in Figure 2-8, has four main functions: to dilute airborne contaminants, to supply breathing oxygen, to protect a facility from large temperature changes, and to provide people with air that is comfortable for working. To illustrate the way an HVAC system acts to serve these functions, consider the operation of the dual-duct air conditioning system shown in Figure 2-8. Outside air is introduced through dampers (see Figure 2-9) which filter the incoming air and control

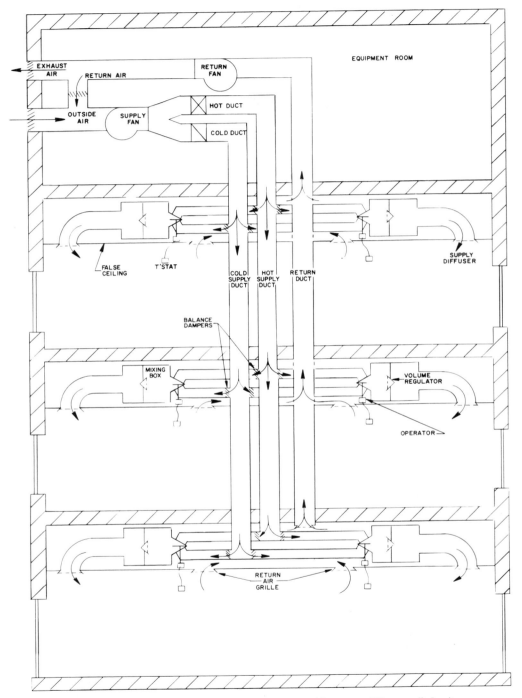

Figure 2–8 Diagram of dual-duct HVAC system. (Courtesy of Honeywell, Inc.)

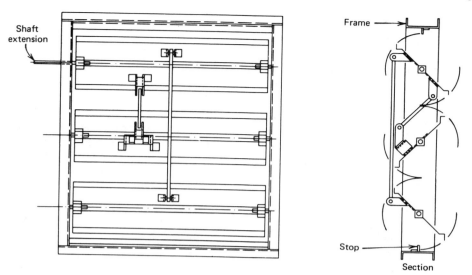

Figure 2-9   Air damper. (From Faye C. McQuiston and Jerald D. Parker, *Heating, Ventilating, and Air-Conditioning,* © 1977. Reprinted by permission of John Wiley & Sons, Inc., New York.)

its quantity. This outside air is mixed with return air (the amount of which is controlled by a return air damper) and blown by a supply fan into both a hot duct and a cold duct. The air in the hot duct passes through heating coils and is sent at a preset temperature to the rooms where it is used. The air in the cold duct goes through cooling coils and thence to the rooms; its temperature is also set at the point where it leaves the cooling coils. At each room, the cold and hot air go into a mixing box controlled by a thermostat; from the mixing box, the air enters the room. Later the air, with heat, cold, and contaminants from the room, is removed through a return air grille and exhausted to the outside and/or returned as part of the intake air.

Although the preceding description applies to a dual-duct system only, its components serve functions which are found in all HVAC systems. Every HVAC system has heat transfer surfaces to enable heating or cooling to take place, every HVAC system has some means of transporting its working fluid to the point of use from the point where heating or cooling is supplied, and every HVAC system has a set of controls which govern its operation. Inspecting each one of these component areas is necessary in a complete energy audit. Since the most common systems are those in which air is the working fluid, the rest of this discussion will be confined to air systems.

A heat transfer surface is a surface where a hot or cold fluid gives up or receives heat from air that is passing around it. Typically, heat transfer surfaces are designed so that a hot or cold liquid flows through pipes surrounded by fins, with the fins used to increase the heat transfer rate. In this part of the HVAC system, points to be inspected are the fluid flow and the heat transfer surfaces. Do relevant gauges show that fluid is flowing? Are "hot water" pipes actually hot? Is fluid flow noise present?

In examining heat transfer surfaces, are the fins and coils clean, or are they fouled with dirt or grease or dust?

The air transportation system moves air from the outside, mixes outside air with return air, and removes used air either to the outside or to the supply fans for use as return air. The main components of this system are the ductwork, dampers, and filters and the fans, blowers, and associated motors. The ductwork can have insulation hanging loose, can leak air through untaped seams, and can be crushed by adjacent piping. The loose insulation can be detected by removing one duct panel and examining the inside with a flashlight; untaped seams and crushed ducts can be detected by a quick visual inspection. Dampers and filters can also be inspected visually. The dampers should be clean and should close, and their controls should be connected; filters should be installed and reasonably clean. Return air grilles should also be inspected to see whether they need to be cleaned.

In addition to ductwork, the fans and blowers should be examined carefully. Fans and blowers should be connected, and the belts should be aligned correctly; i.e., the fan pulley and the motor pulley should be in a straight line. Be particularly thorough in examining fan belts and motor connections; the authors have observed instances where motors had been installed and were running but were not connected to the fans they were supposed to drive. (An easy way to solve this problem is to disconnect the motor.) Motors should also be inspected to see that they are properly connected with balanced voltages to both legs of a three-phase system. Motors should also be free from excess bearing noise. Fans or blowers should be reasonably clean, since accumulated dust detracts from their efficiency, the shaft should not bind, and the fan or blower should rotate in the correct direction.

The control system compares temperatures and pressures with preset values and, depending on the magnitude of the difference, attempts to correct the discrepancy. Controls work by detecting pressure or temperature differences, comparing these with standards, and sending electrical or pneumatic signals to open or close dampers, open or close valves, and turn furnaces, chillers, and blower motors on or off. Clearly it is important that the control system function properly; otherwise, the HVAC system cannot work as intended.

The first step in inspecting the control system is to examine the thermostats. A reliable industrial thermometer (available for $15–$20) can be used to calibrate each thermostat thermometer; if the temperature difference between the two temperature readings is significant, the thermostat should be checked by a vendor. Then the thermostat set temperature should be raised and lowered to see if the heat or cold comes on or shuts off; this is a test of the entire heat control system. (In this process, the person should see whether the room is being cooled or heated more than is necessary. Ideally, a person in a jacket should be cool in the winter, and a person in shirt sleeves should be warm in the summer.) Gauges should be checked to see whether they are connected and whether they are reading within the correct operating range. The compressor that supplies compressed air to a control system should be inspected to see that it is working properly and not leaking oil or water into the control system; if either oil or water gets into the controls, a complete replacement of the control system is often necessary.

2.1.2.4 The electrical system. This system consists of transformers, wiring, switches, and fuses—all the components needed to enable electricity to move from the utility-owned wires at the facility boundary to its point of use within the company. By our definition, this system does not include lights, motors, or electrical controls—lights are discussed in Section 2.1.2.5, motors in Section 2.1.2.8, and controls in Section 2.1.2.3. Most energy problems associated with the distribution of electricity are also safety problems, and solving the energy problems helps to solve the safety problems.

Electricity enters a facility at a transformer. The area around the transformer should be dry, the transformer fins should be free from leaves and debris so that they can perform their cooling function, and the transformer should not be leaking oil. The failure of a transformer to meet any one of these conditions is a matter for concern and should be sufficient justification for a call to the local electrical utility or, if the transformer is company-owned, to the person or department in charge of maintaining the electrical system.

[A more detailed audit of transformers should also include drawing a small (1-pint) sample of transformer dielectric fluid and examining it both visually and for dielectric strength. If the fluid is brown, the dielectric has been contaminated by acid; if it is cloudy, it is contaminated with water. The dielectric strength should be 20,000 v/cm or equivalent. Both the color and the dielectric strength should be recorded for comparison against future readings.]

In examining transformers, also check to see whether any company-owned transformer is serving an area that is not used. A transformer that is connected but not in use is wasting 2–4% of its rated capacity in core losses. These losses can be avoided by disconnecting the transformer or by installing switching between the transformer and the electrical lines from the utility.*

A person performing an energy audit should also examine switch boxes. Danger signs and symptoms of wasted energy include signs of arcing such as burned spots on contacts, burned insulation, arcing sounds, and frayed wire. Other things requiring observation are warm spots around fuse boxes and switches and the smell of warm insulation. Any of these symptoms can indicate a fire hazard and should be checked in more detail immediately.

2.1.2.5 Lights, windows, and reflective surfaces. The functions of this system are to provide sufficient light for necessary work, to enable people to see where they are going, to assist in building and area security at night, to illuminate advertising, and to provide decoration. In examining these functions, the person performing the energy audit should ask, first, whether any light is needed and, second, whether the correct amount of light is being used.

The first place to look is in corridors and at work stations. The range of footcandles for each task is given in Table 2–1. (Lighting is discussed in detail in Chapter 8.)

*Personal communication from Mr. Bryan Drennan, Customer Services, Utah Power and Light Company, Salt Lake City, Utah. Also see Problem 2.4.

TABLE 2-1 RECOMMENDED LIGHTING LEVELS[a]

Building-space type	Guideline illuminance range (fc)[b]
Commercial interiors	
Art galleries	30–100
Banks	50–150
Hotels (rooms and lobbies)	10–50
Offices	30–100
Restaurants (dining areas)	20–50
Stores (general)	20–50
Merchandise	100–200
Institutional interiors	
Auditorium-assembly places	15–30
Hospitals (general areas)	10–50
Labs-treatment areas	50–100
Libraries	30–100
Schools	30–150
Industrial interiors	
Manufacturing areas	50–1000
Ordinary tasks	50
Difficult tasks	100
Highly difficult tasks	200
Very difficult tasks	300–500
Most difficult tasks	500–1000
Exterior	
Building security	1–5
Floodlighting (low-high brightness of surroundings)	5–30
Parking	1–5

[a] Although it is recommended that the reader refer to the *IES Lighting Handbook*, 1981, to obtain the most accurate data on illuminance level selection, this table is included as a guide for initial evaluation of levels found during a building survey. When designing or redesigning a space, refer to the IES tables.

[b] To convert footcandles to lux, multiply value by 10.76.

Source: Reprinted from W. C. Turner, Senior ed., *Energy Management Handbook*, 1982, with permission of John Wiley & Sons, Inc., New York.

An inexpensive ($30–$50) industrial light meter can be used to discover whether the light being used falls within recommended guidelines. Other factors that can make an area appear to be dark are walls, floors, or ceilings that are painted in dark colors or are greasy or dirty. Painting with light colors or cleaning these surfaces more often can make existing light more effective and thereby save money. Major savings can also be obtained by removing lights that are not serving useful functions and by installing timing switches for lights that are turned on at predictable intervals.

Other places where lighting should be examined carefully include warehouses and storage areas. Lights should not be on in these places unless the lights serve some function—illuminating storage areas for assistance in finding product or reading labels, security, or other identified functions. It makes little sense to have a light illuminating an area where people do not need the light, and yet this happens frequently.

While examining this system, windows should also be considered for their value in adding to the light possible in a given area. One problem is that windows admit radiant heat as well as light, and it may be more expensive to remove the heat than to supply the light. If it is decided that the windows are more expense than profit, they should be either painted white (or with some reflective paint), or they should be covered. If they are to be used, they should be kept clean.

Ballasts are also an important part of a lighting system. Each ballast uses 5–20% of the power of the light it is associated with, so that removal of the light should be accompanied by removal of the ballast. (The ballast can be stored for future use, saving additional replacement costs.) Ballasts should also be replaced if they buzz or smoke.

Exterior lighting provides another example where light energy is often wasted. In motels, for example, peripheral lighting is often left on both day and night. Such waste can be easily corrected with a time clock or with a light switch turned on by a photocell and off by a time clock. Each of the perimeter and outside lights should be carefully considered to see whether it needs to be on when it is currently scheduled, whether the amount of light that is now in place is more than is needed for the intended function, and whether more efficient lighting sources would work as well as those now being used. The value of replacing existing lights with new lights depends on the wattage replaced and the annual hours of operation and is easily calculated.

2.1.2.6. The hot water distribution system.

This system distributes hot water for washing, for use in industrial cleaning, and for use in kitchens. Its main components are hot water heaters, storage tanks, piping, and faucets; electric boilers and radiators are also found in some facilities.

In performing an energy audit of this system, first see if more hot water is heated than is needed; second, determine whether the hot water can be heated to a lower temperature without adverse effects; and, third, assess the need for insulation.

To estimate the need for hot water, look at the industrial cleaning uses first and then examine the personnel needs for washing. The amount and temperature of hot water needed in industrial processes will be defined by each process, and the auditor should make sure that neither the amount nor the required temperature are overstated. Each process should be examined to see whether cold water can be substituted for hot. Sometimes a change in detergents is necessary, but often the unavailability of hot water leads to the discovery that cold water works as well, without any other change in procedures. The generally accepted figures for warm water for washing needs are 2–3 gal/day per person in offices, 20–25 in homes, and 40 in hospitals.* If

---

*See *Identifying Retrofit Projects for Buildings* in the Bibliography.

your facility exceeds these amounts, it may be worthwhile to valve off (carefully!) one or more hot water tanks to see if any complaints result. Another necessary step in this audit is to check hot water faucets for leaks. Hot water leaks are expensive: A continuous leak of 1 gal/h of water at 155°F represents a loss of 7 million Btu/year, equivalent to about 7000 ft³ of natural gas.

To find out whether the temperature of the hot water can be lowered, first determine the temperature now. If the water is used for washing, its delivered temperature can be reduced to 105°F by adjusting the thermostat on the hot water heater. If the water is used in a kitchen, it must be hot enough to meet local health codes—usually at least 140°F. In many cases, this temperature is achieved by heating the water to an excessive temperature at the hot water tank, at the cost of increased heat loss between the hot water tank and the kitchen. This loss can be avoided by having water at usual temperature in the kitchen with a booster hot water unit in the kitchen near the dishwasher. (This unit should be on a timer so that water is heated only when needed.) Once the present temperature has been determined, lower the hot water thermostat to the desired level, possibly in 5° increments to avoid complaints. Also consider eliminating hot water completely—most of us have used washrooms in gasoline stations where no hot water was provided without ill effects.

Where the need for hot water has been established, it may be desirable to insulate pipes, hot water tanks, and valves. If a pipe or tank feels hot to the touch, its temperature is probably 140°F or higher, and it is probably losing a significant amount of heat to the surrounding air. (Check the hot water themostat. In one case, both the minimum and the maximum temperatures were set at 180°F, and one resident had complained about scalding.) Valves and flanges have large heat transfer surfaces, and custom-made insulation for them can sometimes be a worthwhile investment.

2.1.2.7 Air compressors and the air distribution system.

Air compressors and the air distribution system provide motive power for tools and some machinery, but their more usual function is to provide air to operate the heating, ventilating, and air conditioning system. If you use compressed air to run equipment, look for leaks and for places where compressed air is allowed to vent into the air. Such leaks can be expensive. One way to determine how much energy is being used in the air distribution system is to attach a recording ammeter to one leg of the motor driving the air compressor. This gives both the time and the amount of energy consumption and can point to excess usage or to leaks as possible problems. The assignment of dollar values to air leaks of various sizes is possible, but the amount of air lost to leakage is usually a small fraction of the total air used, and more attention should probably be paid to the ways in which the air is used. In any case, monitoring the electrical consumption of the compressor motor can determine whether this use of energy is worth the attention of the auditor.

While considering this system, think of replacing air-powered equipment by equipment powered by electricity. An air-powered hoist, for example, uses 5 hp in the air compressor for every hp that would be used if the hoist were electrical.*

*See W. C. Turner, "The Impact of Energy on Material Handling," in the Bibliography.

Where the main function of the air compressor is to control HVAC equipment, water or oil in the control air can wreck the controls on the equipment. Most air compressors have water or oil traps at the bottom, and they can be inspected to see if they are acting as the manufacturer intended. Obtain the operating manual before attempting the inspection; otherwise you may manipulate the safety valve rather than the water trap, with hazardous results.

If water or oil is not removed at the compressor, it may travel to the thermostats and impede their operation. Each thermostat operates differently, however, and inspecting a thermostat for oil or water is a task best left to a vendor or to trained service personnel.

Electric motors transmit power to compressors through belts, and a misalignment of the belt pulleys can cause severe damage and early failure of both bearings and belts. Other problems that can occur with compressors include inoperable switches, gauges that do not work, loose or frayed wiring, and leaks. Look for these problems, and listen for sounds of escaping air. If compressor problems persist, the cost savings of a working control system more than pay for the help of a trained technician.

2.1.2.8 Motors and ovens.

Each manufacturing process has opportunities for energy management, and each offers ways for the unwary to create operating problems in the name of energy management. The best way to avoid such operating problems is to include operating personnel in the energy audit process (see Section 2.1.3) and to avoid rigid insistence on energy conservation as the most important goal. Some energy management techniques that should, however, be considered are reducing process temperatures, rescheduling equipment to reduce peak loads, and waste heat recovery. They are discussed elsewhere in this book.

Most manufacturing facilities have motors and ovens, and they offer ways to save energy and money. Inspect a motor to determine the condition of its bearings and its pulley and belt alignment. The belt alignment has been discussed earlier; the bearing condition can be checked with an industrial stethoscope using the Tornberg procedure (Chapter 14 [13]). In this procedure, applied to motors that use chains or belts to drive equipment, the stethoscope is used to measure the noise at both ends of the motor. In a normal motor, the active end will read 2–3 db louder than the fixed end. If there is no difference or if the fixed end is louder than the active end, the bearings at the fixed end are worn; if the active end is 4–6 db louder then the fixed end, the bearings are badly worn; if the active end is 7–8 db louder, the bearing housing is turned inside the motor; and if the active end is 9 db or more louder than the fixed end, the motor should be replaced immediately. This procedure, developed at Safeway Stores by Mr. Harold Tornberg, has proved practical in industry.

Ovens are also frequently found in manufacturing processes. If an oven uses a significant amount of energy, the energy audit should include an infrared examination of the oven to detect breaks in the refractory, missing insulation, and other points where large temperature differences indicate significant losses of heat. In general, gaskets, refractory, and controls should be checked, and operating procedures should be examined in detail. Worn gaskets leak cold air in and hot air out. Missing refractory and insulation both allow heat to escape from the oven; if enough is missing, the losses can be expensive. If the oven thermostats are out of calibration, two possibilities exist: Either the oven is warmer than intended, and energy can be saved by lowering its temperature, or it has been operating without problems at a lower temperature than originally anticipated, and it should be kept operating at the lower temperature. Operating procedures should be checked to see that the oven door is not open any more often than necessary. (This may be more relevant in a bakery than in a situation where purging of vapors is required.) Oven vendors are also good sources of ways to save energy.

2.1.3 Performing Preliminary Audits

In Section 2.1.2 we described the energy-consuming systems and told what to look for in each system. The next step is to form an energy management committee. Before forming this committee, however, the person leading the energy audit effort should personally use the preceding information to perform two preliminary energy audits to make sure that the committee effort will be worthwhile. The first audit is done under usual working conditions and is aimed at practices that are consistently expensive or wasteful. The second audit (the 2 a.m. audit) is done at midnight or later when energy consumption should be a minimum. The objective of this second audit is to find lighting that is unnecessary, motors that are running but not needed, rooms that are warm but not warming people, air being cooled unnecessarily, and air and steam leaks that might not be detected under the noise conditions of daily operations.

During both audits, a notebook should be kept handy to record problems to be brought to the attention of the audit committees to be formed when the company takes on energy management as a commitment. This commitment can be obtained more easily by the preparation of a list of low-cost, high-return measures, some of which might be presented to the person responsible for company policy in this area. If the auditor cannot find such measures, the energy management program will fail: Why do it if there is no payback? (Note: We know of no company where energy management has not proved cost effective.)

2.2 ACTION

Action is the second step in the three-step procedure for performing an energy audit. In the first step, analysis, energy consumption data were collected and analyzed, the energy-related systems were examined carefully, ideas for improvement were collected and management commitment was obtained. The next step—action—is to obtain company support for the program, to choose goals, and to initiate action.

2.2.1 The Energy Action Team

Once the preliminary audits have uncovered some energy management measures that can save significant amounts of money or can substantially improve production, the next step is to obtain the necessary money and the support of the employees. This can be done with the help of a committee, preferably called something like the *energy action team*. The functions of this committee are given in Table 2-2.

No program will work within a company without employee support, particularly such a program as energy management which seems to promise employee discomfort at no visible increase in production. One function of the energy action committee is thus to give representation to every important political group within the company. For this purpose, the committee must include people from unions, management, and every major group that could hinder the implementation of an energy management plan. The committee must also include at least one person with financial knowledge of the company, a person in charge of the daily operation of the physical plant, and line personnel in each area of the facility that will be affected by energy management. In a hospital, for example, the committee would have to include a registered nurse, a physician, someone from hospital administration, and at least one person directly involved in the operation of the physical plant. In a university, the committee should include a budget officer, at least one department chairperson, a faculty member, a senior secretary, someone from buildings and grounds, and one or more students.

TABLE 2-2 FUNCTIONS OF THE ENERGY ACTION COMMITTEE

1. Create support within the company for energy management.
2. Generate new ideas.
3. Evaluate suggestions.
4. Set goals.
5. Implement the most promising ideas.

In addition to providing representation, a broadly based committee provides a forum for the evaluation of suggestions. The committee should decide on evaluation criteria as soon as possible after it is organized. These criteria should include first cost, estimated payback period or (for projects with a payback period longer than 2 years) the constant-dollar return on investment (see Chapter 4), effects on production, the effects on acceptance of the entire program, and any mitigating effect on problems of energy curtailment.

The committee has the additional duty to be a source of ideas. These ideas can be stimulated by a detailed energy audit along a route chosen in advance to show problems and areas for improvement. The person leading the audit should be aware, however, that most maintenance personnel become quickly defensive and that their cooperation, and hopefully their support, may be important.

In addition to performing walk-through audits, the specific tasks of this committee are to set goals, implement changes, and monitor results. At least three different kinds of goals can be identified. First, performance goals, such as a reduction of 10% in Btu/unit product, can be chosen. Such goals should be modest at first so that they can be accomplished—in general, 10-30% reduction in energy usage for companies with little energy management experience and 8-15% for companies with more. These goals can be accompanied by goals for the reduction of projected energy costs by a similar amount. The more experienced companies generally have fewer easy saving possibilities, and lower goals are more realistic.

A second type of goal that can be established is an accounting goal. The ultimate objective in an energy accounting system is to be able to allocate the cost of energy to a product in the same way that other direct costs are allocated, and this objective guides the establishment of preliminary energy accounting goals. A preliminary goal would therefore be to determine the amount of electricity and the contribution to the electrical peak from each of the major departments within the company. This will probably require some additional metering, but it has been the authors' experience that such metering pays for itself in energy saving (induced by a better knowledge of the energy consumption patterns) in 6 months or less.

The third type of goal is that of employee participation. Even if an energy management program has the backing of the management, it will still fail without the support and participation of the employees. Ways to measure this include the number of suggestions per month; the dollar value of improvements adopted as a result of employee suggestions, per month; and the number of lights left on or machines left running unnecessarily, on a spot inspection. Work sampling has been used to estimate the percentage of time that people are working at various tasks—it can be used equally well on machines.

In addition to providing and evaluating ideas, setting goals, and establishing employee support, the energy action committee has the additional duty of implementing the most promising ideas that have emerged from the evaluation process. Members of the committee have the responsibility to see that people are assigned to each project, that timetables are established, that money is assigned, and that progress reporting procedures are set up and followed. It is then the committee responsibility to follow up on the progress of each project; this monitoring process is described in detail in Section 2.3.

2.3 MONITORING

Management is not complete without monitoring and its associated feedback, and neither is the energy audit process. In an energy audit, monitoring discloses what measures contributed toward the company goals, what measures were counterproductive, and whether the goals themselves were too low or too high.

Monitoring consists of collecting and interpreting data. The data to collect are defined by the objectives chosen by the committee described. At the very least, the electrical and gas bills (and those of other energy sources, where relevant) are examined and their data graphed each month. Monthly graphs should include the total energy used of each type (kwh of electricity and therms (10^5 Btu) of gas), the peaks (if they determine part of the cost of electricity or gas), and any other factors that contribute to the bills. At the same time, other output-related measures, such as Btu/ton, should also be calculated, recorded, and graphed.

The data monitored should be related to the lowest organizational level possible so that they provide feedback as directly as possible. Often this requires that recording instruments be installed in a number of departments in addition to the meters required by the utility company. The additional expense is justified by increased awareness of the timing and amounts of energy consumed, and usually this awareness leads to a reduction in energy costs. Metering at each department also enables management to determine where the energy is consumed and, possibly, what is causing the energy consumption. Such metering also helps each department manager to control the consumption of his or her own department.

The results of monitoring are actions. Find what is good, and copy it elsewhere. Find what is bad, and avoid it elsewhere. If goals are too high, lower them. If goals are too low, raise them. Wherever the difference between planned objectives and achievements is great, initiate an analysis to determine the reason and then develop objectives, initiate action, and monitor the results. In this way, the analysis, action, and monitoring process repeats itself.

2.4 SUMMARY

In this chapter, the mechanics and reasoning behind an energy audit have been explained. First is the analysis of past data, then the examination of each energy system, then the formation of an energy action committee, and last the necessity and mechanics of monitoring.

BIBLIOGRAPHY

BAUMEISTER, THEODORE, EUGENE A. AVALLONE, and THEODORE BAUMEISTER III, *Marks' Standard Handbook for Mechanical Engineers,* 8th ed., McGraw-Hill, New York, 1978.

HONEYWELL, INC., *Variable Air Volume Systems Manual,* 1980 ed., Manual No. 74–5461, Honeywell, Inc., Minneapolis, 1980.

Identifying Retrofit Projects for Buildings, Federal Energy Management Program, FEA/D–76, Washington, D.C., Sept. 1976.

IES Lighting Handbook, Illuminating Engineering Society of North America, New York, 1981.

Industrial Ventilation: A Manual of Recommended Practice, 15th ed., American Conference of Governmental Industrial Hygienists, Lansing, Mich., 1979.

KENNEDY, W. J., Jr., "How to Conduct an Energy Audit," in *Proceedings of the 1978 Fall AIIE Conference,* Institute of Industrial Engineers, 25 Technology Park/Atlanta, Norcross, Ga., 1978.

KENNEDY, W. J., Jr. "An Introduction to HVAC Systems," in *Proceedings of the 1980 Fall Industrial Engineering Conference,* Institute of Industrial Engineers, Norcross, Ga., 1980.

MCQUISTON, FAYE C., and JERALD D. PARKER, *Heating, Ventilating, and Air-Conditioning,* Wiley, New York, 1977.

PETTERSON, CARL, *Energy Survey and Conservation Manual,* Granite School District, Salt Lake City, 1980.

TURNER, WAYNE C., "The Impact of Energy on Material Handling," *Proceedings of the 1980 Conference on the Materials Handling Institute,* Material Handling Institute, Pittsburgh, Pa., 1980.

TURNER, WAYNE E., Senior ed., *Energy Management Handbook,* Wiley-Interscience, New York, 1982.

1977 Fundamentals Handbook and Product Directory, American Society of Heating, Refrigerating, and Air-Conditioning Engineers, Inc., New York, 1977.

1978 Applications Handbook and Product Directory, American Society of Heating, Refrigerating, and Air-Conditioning Engineers, Inc., New York, 1978.

1979 Equipment Handbook and Product Directory, American Society of Heating, Refrigerating, and Air-Conditioning Engineers, Inc., New York, 1979.

1980 Systems Handbook and Product Directory, American Society of Heating, Refrigerating, and Air-Conditioning Engineers, Inc., New York, 1980.

QUESTIONS

2.1. Which performance measure should be used in setting up an audit procedure for a series of buildings: Btu/ft²/year or Btu/year? Why?

2.2. Select a building for study. Perform an energy audit on that building. Collect bills for a 2-year period, walk through the building, and choose at least 10 different measures for possible implementation. Estimate possible savings from each measure.

2.3. In Problem 2.3, what additional benefits can be achieved by reballasting with more efficient ballasts?

2.4. Outline a procedure to use in auditing a company's use of ovens. Include both maintenance and scheduling aspects.

PROBLEMS

2.1. The following data describe energy consumption for the elementary schools in the Blizzard School District for 1 year. Use these data to help develop energy standards for the Blizzard School District. Specifically isolate the 6 best and the 6 worst schools, and outline how you would use all 21 schools in the development of an energy audit program.

TABLE FOR PROBLEM 2.1

School	Natural gas consumption (Mcf)	Electrical consumption (kwh)	Area (ft²)
Acadia Park	7240	316,720	52,300
Cotton	3489	147,280	44,896
Crestview	3098	117,420	51,400
Crim	4086	163,860	42,800
Droggs	6700	445,920	63,300

TABLE FOR PROBLEM 2.1 (*CONT.*)

School	Natural gas consumption (Mcf)	Electrical consumption (kwh)	Area (ft^2)
East Mill	4360	98,640	41,937
East Wood	6028	197,910	38,000
Edwards	4794	133,340	42,800
Farnsforth	3897	380,640	57,600
Fremont	4800	216,160	46,400
Hillsdale	4721	147,680	52,400
Hill View	6496	198,480	49,200
Lake Ridge	4047	130,140	57,200
Lincoln	4655	107,620	37,500
Madison	3757	75,480	37,100
Magna	2139	74,560	36,700
Milltown	3146	89,800	39,900
Monroe	6062	564,360	58,626
Moss	3737	311,040	57,600
Oak Hills	4139	99,120	49,000
Oakwood	4770	167,360	45,900

2.2. An outlying building has a 25-kw company-owned transformer that is connected all the time. A call to a local electrical contractor indicates that the core losses from comparable transformers are approximately 3% of rated capacity. Assuming that the marginal electrical costs are $.10/kwh and $10.00/kw of peak demand, that the average building use is 10 h/month, and that the average month has 730 h, estimate the annual cost saving from installing a switch that would energize the transformer only when the building is being used.

2.3. A large facility you have just inspected has 350 two-lamp fluorescent fixtures, with each fixture using 253 w. New improved ballasts can save 48 w/fixture. The facility operates with one 8-h shift each day, 5 days/week, 50 weeks/year. The demand peak is during the daytime. Present electrical costs are $.10/kwh/month, with a demand charge of $4.50/kw/month. What is the simple payback period on this proposed investment?

2.4. Your company receives electrical power through company-owned transformers. Your annual payroll is $10 million. The usual working year is 250 days. If one of your transformers fails, the minimum time to repair it is 2 days, with employees paid for that time and with an equal amount of time needed to make up the lost production. A contractor has offered to perform an inspection and maintenance service, correcting any defects found, for $600 per inspection. If the transformers are inspected once each year, the probability of detecting and correcting incipient transformer problems is .95. If the inspection is performed twice each year, the probability is .98; if quarterly, .99; and if every other year, .90. How often should the transformers be inspected?

2.5. In examining a large walk-in electric oven operating at 1000°F, you note that the seal around the door is deteriorating and that there is a draft of hot air. Measurement shows this to have a volume of 80 cfm. The electric rates have a lowest-block cost of $.065/kwh with a peak power rate of $7.50/kw/month. This oven is operated 10h each day, 5 days each week, during peak demand hours. Your HVAC system efficiency is estimated as 70%. Ambient room temperature is 80°F. How much is the 80-cfm leak adding to the monthly electric bill paid by the company?

2.6. In inspecting a school, you have counted 15 broken windows. The average size of the window holes is about 8 × 10 in. Measurements indicate that the average wind velocity through the holes is about 6 ft/s. The vicinity of the school has 6100 heating degree days per year, and gas costs $.65/therm. Assuming an 80% efficient gas heating unit, how much are the broken windows costing the school each year? Assume that humidity adds .013 lb water vapor per lb of dry air and that the specific heat of water vapor is .489 Btu/lb•°F.

Chapter 3

Energy Sources
and Rate Structures

3.0 INTRODUCTION

Energy costs have been increasing and supplies or proven reserves dwindling as demonstrated in Chapter 1. The impact of these increases is vividly demonstrated in the rate schedules billing for the various fuel sources, yet few managers take the time to peruse and understand their billing procedures.

Why? The reasons are many, but the main ones appear to be the following:

- Rate schedules are sometimes very complicated. They are difficult to understand, and the explanations developed by state utility regulating boards and utilities themselves often confuse rather than clarify.
- Energy is an overhead item. Even though energy is often a substantial portion of the product's cost, it is almost always treated as an overhead item rather than a direct cost. This means it is more difficult to account for and control. Consequently, management does not give energy cost the attention it deserves.

This must change and is changing. More and more managers are trying to understand their rate schedules, and sometimes they even participate in the rate schedule hearings.

Soon managers will be able to say what electric rate schedule they are under and how much they are billed for demand, consumption, power factor, sales tax, etc. This chapter is designed to help present and future managers learn about rate schedules.

The chapter covers rate schedules for the major energy sources utilized in this country. First, a considerable amount of time is devoted to electricity. Then attention is given to gas, fuel oil, coal, and steam-chilled water.

3.1 ELECTRICITY

Perhaps the best way to understand utility billing for electricity is to examine the costs faced by the utility itself. The major cost categories are the following:

- *Physical plant.* This is often the single biggest cost category. The larger and higher technology plants required and the increasing cost of necessary pollution control mean the physical plant cost is still growing in relativity. A most important point to grasp here is that the utility *must* have sufficient capacity to supply the *peak* needs of its customers and have some equipment in reserve in case of equipment failures. Otherwise insufficient energy will occur at times, leading to brownouts or even blackouts. This is why so many utilities are urging their customers to reduce their peak demand.

- *Transmission lines.* Another major cost category is that of transmission lines to carry the electricity from where it's produced to the general area of where it's needed. Transmission is done at relatively high voltages to minimize I^2R (resistance) losses. Relatively, this cost can be large or small depending on the distances involved.

- *Substations.* Once the electricity reaches the general area where it's needed, the voltages need to be reduced to levels where the electricity can be accepted by the customer. This is done with step-down transformers at substations.

- *Distribution systems.* The power voltage electricity must now be delivered to the individual customers through the local distribution system. Some customers (e.g., large industrial users) may tap directly into the transmission lines, while some (e.g., residential users) may be at the end of a four- or five-step-down sequence. Distribution system costs include utility poles, lines, transformers, and capacitors.

- *Meters.* Although relatively small, meter costs are really a separate item as they form the interface between the utility company and customer. The cost can range from under $50 for a residential meter to $1500 or more for an industrial customer requiring consumption demand and power factor information (these will all be discussed later).

- *Administrative.* Administrative costs include executive salaries, middle management salaries, technical and office staff salaries, taxes, insurance, equipment maintenance, and interest on debt. This cost category can be quite large. For example, the interest on debt for a large nuclear power plant costing over $1 billion can be substantial. In addition, enough revenue must be generated to provide a reasonable profit to stockholders or a reasonable rate of return to bond holders.

- *Energy.* Once the power plant, distribution lines, etc., are all in place, some energy must be purchased to generate the electricity. Costs can be small (e.g., hydroelectric plants) or quite significant (e.g., fuel oil plants). Of course, fossil fuel electric plants have experienced a dramatic increase in fuel cost recently.

The next step is to determine how these costs are to be allocated to the various customers. It is important to recognize that the billing procedure (rate schedule) should be designed to reflect true costs so the customers can understand the problems of the utilities and can help the utilities minimize these costs. Recent rate schedules and proposed new ones do a much better job of this than has been done in the past, but more needs to be done.

There are several classes of customers that an electric utility must serve. These classes vary in complexity of utilization, amount of consumption, and priority of need. Consequently, the various state regulating agencies and utilities have developed several levels of rate schedules. Figure 3–1 presents a generalized breakdown of these levels. (Some of the terms in Figure 3–1 may be unfamiliar to you at this point, but we shall remedy that shortly.)

3.1.1 Residential Rate Schedules

As discussed in Figure 3–1, there are many residential users, but each is a relatively small consumer. Peaking of demand for residential users at the same time is a problem, but thus far few utilities have managed to do much about it in terms of rate schedules. An exception is charging more for energy during peaking months, but this still leaves the daily peaking problem. A typical residential bill, therefore, charges only for the following:

- *Administrative:* This fee is to cover the fixed cost of serving the custmer. It is a flat fee per customer regardless of the amount consumed.
- *Energy charge:* This is to cover the actual amount of electricity consumed measured in kilowatt-hours.

Level	Comments	Typical schedule bills for		
		Consumption (kwh)	Demand (kw)	Power factor (kvar)
1. Residential	Small user but large numbers of them	✓		
2. Commercial	Small to moderate user; relatively large numbers	✓		
3. Small industrial	Small to moderate user; fewer customers	✓	✓	
4. Large industrial	Large user with low priority; typically, only a few customers in this class, but they consume a large percentage of the electricity produced	✓	✓	✓

Figure 3-1   Generalized breakdown of electric rate schedules.

- *Fuel cost adjustment:* The energy charge is based on an average cost for the fuel consumed (natural gas, fuel oil, coal, etc.). If the utility has to pay more than this average for the fuel, the cost is "passed on" to the customer through use of a prescribed formula. The fuel adjustment is often a substantial proportion of the bill.

Figure 3-2 presents a typical rate schedule for a residential customer.

Customer charge: $4.42/customer/month

Energy charge: On-peak season (June through October):
                        All kwh @ 5.200¢/kwh

                Off-peak season (November through May):
                        First 600 kwh/month @ 4.943¢/kwh
                        All additional kwh/month @ 3.709¢/kwh

Fuel adjustment: (A formula is provided by the utility to calculate the fuel adjustment charge. It is rather complex and will not be covered here.)

Figure 3-2 Typical residential rate schedule. (Courtesy of Oklahoma Gas and Electric Co.)

Several items are noteworthy in this schedule. First, this utility has chosen to attack the peaking problem (see the discussion earlier in Section 3.1) by charging more for electricity consumed in the months when peaks occur (summer months in this case).

Second, during the peak season there is a constant charge for all energy (5.2 cents/kwh) regardless of the amount consumed. In the off-peak season, however, a *declining block* approach is used. A declining block schedule charges less for the next increment of energy as more energy is used. This is a logical approach and one that is very popular today.

Another approach is an *increasing block* where more is charged per increment as the consumption level increases. This would tend to minimize waste, but it is against cost of service rationale and is not popular today. The *lifeline* approach is an attempt to provide a subsistence level of electricity at a very reasonable rate so those on welfare and/or retirement incomes can meet their basic energy needs. A typical low-use residential service rate is shown in Figure 3-3. This schedule is an attempt to meet the needs of those on fixed incomes.

Other approaches being examined and sometimes used for residential customers include the following:

- *Time of day pricing.* To handle the daily peaking problem, more is charged for energy consumed during peaking times. This requires relatively sophisticated meters and is not very popular today for that reason. It is becoming more popular, however, and will likely continue to become so.
- *Peak shaving.* Some utilities offer a discount to residential customers if the utility can hook up a remote control unit to cycle large electricity users in the

home (usually air conditioners). This way the utility can cycle air conditioners on and off periodically to help shave demand. Apparently, there is little to no discomfort experienced by the customer since the cycling is done over short periods of time. This approach is rapidly gaining in popularity.

This schedule is appropriate for all customers whose monthly consumption never exceeds 500 kwh and whose consumption exceeds 400 kwh no more than twice in a year.

Customer charge: $4.42/customer/month

Energy charge: 4.132¢/kwh

Fuel adjustment: (A formula is provided for calculating this charge.)

Figure 3-3   Low-use residential schedule. (Courtesy of Oklahoma Gas and Electric Co.)

3.1.2 General Service Rate Schedules

A general service rate schedule is used for commercial and small industrial users. It is a simple schedule usually involving only consumption (kwh) charges and customer charges. Sometimes, capacity (demand) charges are used, but this requires a demand meter, so it is not too prevalent. (See Section 3.1.3 for a detailed discussion of demand charging.)

The energy charge is usually substantially higher than for residential users for various noneconomic reasons. For example, one rate schedule charges almost 8 cents/kwh for commercial users during peak season but only a little more than 5 cents/kwh for residential users during the same season.

3.1.3 Small Industrial Rate Schedules

A small industrial rate schedule is appropriate for small industrial users and perhaps large commercial ones. The service becomes more complex and the consumption higher. Consequently, the billing becomes more sophisticated. Usually, the same cost categories occur as in the simpler schedules, but there are additional ones. Some are outlined below.

3.1.3.1 Voltage level.   One degree of complexity is introduced according to what voltage level is needed by the customer. If the customer is willing to accept the electricity at high transmission levels (usually 50,000 v or higher) and do the necessary transforming to usable levels in house, then the utility saves considerable expense and can charge less. If the customer needs the service at a lower voltage, then the utility needs to install transformers and maintain them. The cost of service goes up and so does the bill.

This can be handled in the rate schedule in several ways. One is for the utility to offer a percentage discount on the electric bill if the customer owns its own primary transformer (thereby accepting service at a higher voltage). Another is to charge more for energy as the voltage level decreases. (This is used in a later example; see Figure 3–6.)

Often this is a primary cost-cutting opportunity for industrial users and should be explored. Maintaining transformers is a relatively simple (though potentially dangerous) task, but standby transformers might be needed to avoid costly shutdowns.

3.1.3.2 Demand billing.

Before presenting a small industrial rate schedule example, it is important we understand the concept of demand billing. Consider Figure 3–4 where energy demands on a utility are plotted vs. time for two hypothetical companies. Since the instantaneous demand (kw) is plotted over time, the integration of this curve (i.e., the area under the curve) is the total energy (kwh) consumed (see the shaded area). Furthermore, company B's peak demand equals company A's average demand, and B's peak demand equals its average demand, so the total energy consumed by B equals that of A. Since the kw hours consumed by each are equal, their bills for consumption will be equal, but this seems unfair. Company B has a very flat demand structure so the utility can gear up for that level of service with high-efficiency equipment. Company A, however, requires the utility to be able to supply about twice the demand of B but only for one short period of time. This means the utility must gear up equipment to run only a short period of time, which is quite expensive.

To properly charge for this and to encourage company A to reduce its peak demand, an electric utility will normally charge industrial users for the peak demand incurred during a billing cycle, normally a month. Often substantial cost reductions can occur by simply reducing peak demands and still consuming the same amount of electricity. A good example of this would be to move the use of an electric furnace from peaking times to nonpeaking times (maybe second or third shifts). This means the same energy could be used at less cost since the demand is reduced. Peak shaving will be discussed further in a later chapter.

Demand is measured by a demand meter. It is normally calculated as the

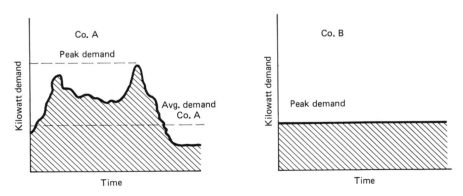

Figure 3–4   Demand profiles for two hypothetical industrial firms.

average demand over a time window (5–30 min) so short-term instantaneous peaks don't normally affect demand bills.

3.1.3.3 Power factor.
Power factor is a complex subject to explain, but it can be a vitally important element in a company's electrical bill. One company the authors worked with had a power factor of 51%. With their billing schedule, this meant they were paying a penalty of 56.9% on demand billing (we'll discuss power factor billing shortly). With the addition of capacitors, this penalty could have been avoided or minimized.

The current required by induction motors, transformers, fluorescent lights, induction heating furnaces, resistance welders, etc., is made up of three types of current:

1. *Power-producing current* (working current or real power). This is the current which is converted by the equipment into useful work, such as turning a lathe, making a weld, or pumping water. The unit of measurement of the power produced is the kilowatt (kw).

2. *Magnetizing current* (wattless or reactive current). This is the current which is required to produce the flux necessary for the operation of induction devices. Without magnetizing current, energy could not flow through the core of a transformer or across the air gap of an induction motor. The unit of measurement of magnetizing current is the kilovar (kvar) or kilovoltamperes reactive.

3. *Total current* (apparent power). This is the current that is read on a voltmeter and an ammeter in the circuit. It is generally made up of magnetizing current and power-producing current. The unit of measurement of total volt-amperes is the kilovoltampere (kva). Most ac power systems require both kilowatts and kilovars.

Power factor is the ratio of actual power being used in a circuit, expressed in watts or kilowatts, to the power drawn from the line, expressed in kilovoltamperes. The relationship of kw, kvar, and kva in an electrical system can be illustrated by scaling vectors to represent the magnitude of each quantity, with the vector for kvar at a right angle to that for kw (Figure 3–5). When these components are added *vectorially,* the resultant is the kva vector. The angle between the kw and kva vectors is known as the *phase angle*. The cosine of this angle is the power factor and equals $\frac{kw}{kva}$.

Since a kw meter does not recognize the reactive power, a company with a low power factor would be billed the same as a company with a high power factor unless some way of billing for low power factor is incorporated into the rate schedule. Most utilities do build in a power factor penalty for industrial users. However, the way of billing varies widely. Following are some of the more common ways.

- Billing demand is measured in kva instead of kw. A look at the triangle in Figure 3–5 shows that as the power factor is improved, kva is reduced, providing a motivation for power factor improvement.

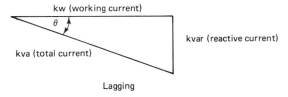

θ = phase angle = measure of net amount of inductive reactance in circuit

cos θ = PF = ratio of *real power* to *apparent power*

kva = $\dfrac{kw}{\cos\theta}$ = $\dfrac{kw}{PF}$ = $\sqrt{(kw)^2 + (kvar)^2}$

Figure 3-5 Diagram of ac component vectors.

- Billing demand is modified by a measure of the power factor. Some utilities will increase billed demand 1% for each 1% the PF is below a designated base. Others will modify demand as follows:

$$\text{billed demand} = \text{actual demand}\left(\frac{\text{base power factor}}{\text{actual power factor}}\right)$$

This way, if the actual is below the base, the billed demand is increased. If the actual is above the base, some utilities will allow the fraction to stay, thereby providing a reward instead of a penalty. Some will run the calculation only if actual PF is below base PF.
- The demand or consumption billing schedule is changed according to the power factor. Some utilities will change the schedule for both demand and consumption according to the power factor.
- Charge per kvar is used. Some companies will charge for each kvar used above a set mimimum. This is direct billing for the power factor.

3.1.3.4 The rate schedule.   The previous few paragraphs were necessary in order to be able to present the rate schedule itself in understandable terms. With all these complex terms and relationships, it's no wonder many managers don't bother to understand their bills. Now, however, you are ready to tear into a typical rate schedule. Consider Figure 3–6. Let's examine the different components.

- *Voltage level.* This utility has chosen to encourage company-owned primary transformers by offering a cheaper rate for both demand and consumption if the company accepts service at a higher voltage level. To analyze the savings for primary transformer ownership, a company simply needs to calculate the energy dollar savings from accepting service at a higher level and compare that savings to the cost of the necessary transformers and annual maintenance thereof. Transformer losses must be absorbed by the company, and the company must provide a standby transformer or make other arrangements in case of a breakdown. (See Problem 3.5 at the end of this chapter for an example.)

Effective in: All territories served

Availability: Power and light service. Alternating current. Service will be rendered at one location at one voltage. No resale, breakdown, auxiliary, or supplementary service permitted.

Rate:

A.   Transmission service (service level 1):

Customer charge: $483.00/bill/month

Demand charge applicable to all kw/month of billing demand:

On-peak season: $341.00 for first 75 kw or less
                              plus $3.76/kw for all additional kw
Off-peak season: $215.00 for first 75 kw or less
                               plus $2.50/kw for all additional kw

Energy charge: 2.503¢ applicable to all kwh/month

B.   Distribution service (service levels 2, 3, and 4):

Customer charge: $204.00/bill/month

Demand charge applicable to all kw/month of billing demand:

On-peak season: $401.00 for first 75 kw or less
                              plus $4.45/kw for all additional
Off-peak season: $253.00 for first 75 kw or less
                               plus $2.95/kw for all additional

Energy charge: 2.671¢ applicable to all kwh/month

C.   Secondary service (service level 5):

Customer charge: $115.00/bill/month

Demand charge applicable to all kw/month of billing demand:

On-peak season: $449.00 for first 75 kw or less
                              plus $4.97/kw for all additional kw
Off-peak season: $283.00 for first 75 kw or less
                               plus $3.29/kw for all additional kw

Energy charge: 2.765¢ applicable to all kwh/month

Definition of season:

On-peak season: Revenue months of June–October of any year.

Off-peak season: Revenue months of November of any year through May of the succeeding year.

Late payment charge: A late payment charge in an amount equal to one and one-half per cent ($1\frac{1}{2}$%) of the total amount due on each monthly bill as calculated under the above rate will be added if the bill is not paid on or before the due date stated on the bill. The due date shall be twenty (20) days after the bill is mailed.

Minimum bill: The minimum monthly bill shall be the customer charge plus the applicable demand charge, as computed under the above schedule. The company shall specify a larger minimum monthly bill, calculated in accordance with the company's allowable expenditure formula in its terms and conditions of service on file with and approved by the commission, when necessary to justify the investment required to provide service.

Determination of maximum demand: The consumer's maximum demand shall be the maximum rate at which energy is used for any period of 15 consecutive min of the month for which the bill is rendered as shown by the company's demand meter. In the event a consumer taking service under this rate, has a demand meter with an interval greater than 15 min, the company shall have a reasonable time to change the metering device.

Figure 3-6 Typical small industrial rate schedule. (Courtesy of Oklahoma Gas and Electric Co.)

<u>Determination of billing demand:</u> The billing demand upon which the demand charge is based shall be the maximum demand as determined above corrected for the power factor, as set forth under the power factor clause, provided that no billing demand shall be considered as less than 65% of the highest on-peak season maximum demand corrected for the power factor previously determined during the 12 months ending with the current month.

<u>Power factor clause:</u> The consumer shall at all times take and use power in such manner that the power factor shall be as nearly 100% as possible, but when the average power factor as determined by continuous measurement of lagging reactive kilovoltampere hours is less than 80%, the billing demand shall be determined by multiplying the maximum demand, shown by the demand meter for the billing period, by 80 and dividing the product thus obtained by the actual average power factor expressed in percent. The company may, at its option, use for adjustment the power factor as determined by test during periods of normal operation of the consumer's equipment instead of the average power factor.

<u>Fuel cost adjustment:</u> The rate as stated above is based on an average cost of $1.60/million Btu for the cost of fuel burned at the company's thermal generating plants. The monthly bill as calculated under the above rate shall be increased or decreased for each kwh consumed by an amount computed in accordance with the following formula:

$$FA = A \times \frac{B}{10^6} \times C + \frac{P}{S}$$

where FA = fuel cost adjustment factor (expressed in dollars per kwh) to be applied per kwh consumed

     A = weighted average Btu/kwh for net generation from the company's thermal plants during the second calendar month preceding the end of the billing period for which the kwh usage is billed

     B = amount by which the average cost of fuel per million Btu during the second calendar month preceding the end of the billing period for which the kwh usage is billed exceeds or is less than $1.60/million Btu; any credits, refunds, or allowances on previously purchased fuel, received by the company from any source, shall be deducted from the cost of fuel before calculating B each month

     C = ratio (expressed decimally) of the total net generation from all the company's thermal plants during the second calendar month preceding the end of the billing period for which the kwh usage is billed to the total net generation from all the company's plants including hydrogeneration owned by the company, or kw produced by hydrogeneration and purchased by the company, during the same period

     P = cost of power purchased by the company from a cogeneration or small power production facility calculated on the basis of the buy-back rate established and approved by the commission times the total kwh purchased from such facility or facilities during the second calendar month preceding the end of the billing period for which the kwh usage is billed

     S = total kwh sales by the company during the second calendar month preceding the end of the billing period for which the kwh usage is billed

<u>Franchise payment:</u> Pursuant to Order No. 110730 and Rule 54(a) of Order No. 104932 of the Corporation Commission of Oklahoma, franchise taxes or payments (based on a percent of gross revenue) in excess of 2% required by a franchise or other ordinance approved by the qualified electors of a municipality, to be paid by the company to the municipality, will be added pro rata as a percentage of charges for electric service, as a separate item, to the bills of all consumers receiving service from the company within the corporate limits of the municipality exacting the said tax or payment.

<u>Transmission, distribution, or secondary service:</u> For purposes of this rate, the following shall apply:

<u>Transmission service</u> (service level 1), shall mean service at any nominal standard voltage of the company above 50 kv where service is rendered through a direct tap to a company's transmission source.

<u>Distribution service</u> (service levels 2, 3, and 4), shall mean service at any nominal standard voltage of the company between 2000 v and 50 kv, both inclusive, where service is rendered through a direct tap to a company's distribution line or through a company-numbered substation.

Figure 3–6 (*Cont.*)

Secondary service (service level 5), shall mean service at any nominal standard voltage of the company less than 2000 v or at voltages from 2 to 50 kv where service is rendered through a company-owned line transformer.

If the company chooses to install its metering equipment on the load side of the consumer's transformers, the kwh billed shall be increased by the amount of the transformer losses calculated as follows:

   1% of the total kva rating of the consumer's transformers times 730 h

Term: Contracts under this schedule shall be for not less than 1 year, but longer contracts subject also to special minimum guarantees may be necessary in cases warranted by special circumstances or unusually large investments by the company. Such special minimum guarantees shall be calculated in accordance with the company's allowable expenditure formula in its terms and conditions of service on file with and approved by the commission.

Figure 3–6 (*Cont.*)

- *Demand billing.* This utility has chosen to emphasize demand leveling by assessing a rather heavy charge for demand. (Actually, charges are regional. For the southwest, this is a rather large charge. For the northeast, it would be cheap.) Furthermore, the utility has emphasized demand leveling during the peaking season (summer in this case).

- *Consumption.* This utility has decided to stick with a flat charge per kilowatt-hour for consumption levels as opposed to a declining block approach. This is a new approach but there appears to be a trend in that direction.

- *Power factor.* The utility has chosen to charge for the power factor by modifying the demand charge (see the section labeled "power factor clause"). They have decided all customers should aim for a power factor of at least 80% and should be penalized for power factors of less than 80%. To do this, the peak demand is multiplied by a ratio of the base power factor (80%) to the actual power factor if the actual power factor is below 80%; there is no charge or reward if the power factor is above 80%:

$$\text{billed demand} = \text{actual demand} \left(\frac{\text{base power factor}}{\text{actual power factor}} \right)$$

where the base power factor = .80.

- *Ratchet clause.* In the previous discussion on demand charges (Section 3.1.3.2), we did not discuss what's come to be known as a *ratchet clause*. To understand the motivation behind ratchet clauses, one must realize that if the utility must supply a peak load in July, it cannot sell that equipment for the other 11 months. Instead, the equipment must be kept and maintained for the next time when it will be needed. To charge for this cost (and to encourage demand leveling over the months), most utilities install a ratchet clause (see the section labeled "determination of billing demand"). In this case, the utility has decided to say that the billed demand for any month is "65% of the highest on-peak season maximum demand corrected for the power factor" of the previous 12 months *or* the actual demand corrected for power factor whichever is greater. For a company with a large seasonal peaking nature, this can be a real problem. A peak can be set in July (air conditioning) that the company in effect pays for for a full year. The impact of ratchet clauses can be significant, but often a company never realizes this has occurred.

- *Miscellaneous.* Other items appearing in the rate schedule include fuel cost adjustment, late payment charge, and minimum bill. The fuel cost adjustment is based on a formula and can be quite significant. Anytime the cost of energy is calculated, the fuel cost adjustment should be included. Late payment charges and minimum bills usually aren't a problem.

- *Sales tax.* One item not mentioned in the schedule that can be quite important is sales tax. With many localities having sales taxes of 6% or more, this can be a significant cost factor. The cost of electrical service should show this. One word of caution: Some states have laws stating that *energy used directly in production should not have sales tax charged to it*. This is important in any industry with energy going to production. Some submetering may be necessary, but the cost savings often justifies this. For example, electricity used in a process furnace should not be taxed, but electricity running the air conditioners would be taxed.

3.1.3.5 Example calculation.   For an example of rate schedule calculations, let's calculate the bill for the company shown in Figure 3–7 using the schedule shown in Figure 3–6.

Month: September 1984

Actual demand: 250 kw

Consumption: 54,000 kwh

Previous high demand: 500 kw (July 1984) (last 12 months)

Power factor: 75%

Service level: Secondary (PLS, service level 5)

Sales tax: 6%

Fuel cost adjustment: 1.15¢/kwh

Figure 3-7 Parameters for rate schedule calculations. (Courtesy of Acme Mfg. Co.)

As a first step, the demand should be calculated:
Power factor correction:

$$\text{billed demand} = (\text{actual demand})\left(\frac{.80}{\text{PF}}\right)$$
$$= 250 \text{ kw}\left(\frac{.80}{.75}\right)$$
$$= 266.67 \text{ kw}$$

$$\text{minimum billed demand (ratchet clause)} = (500 \text{ kw}) (.65)$$
$$= 325 \text{ kw}$$

$$\text{billed demand} = \max. (250 \text{ kw}, 325 \text{ kw})$$

$$\underline{\text{billed demand} = 325 \text{ kw}}$$

demand charge (on-peak season) = first 75 kw   = $ 449.00
                                  (325 kw - 75 kw)($4.97/kw)   = $1242.50

                         total demand charge = $1691.50

Consumption charge:

                              (54,000 kwh)($.02765/kwh)   = $1493.10
                              + fuel adj.(54,000 kwh)(.0115) = $ 621.00

                              total consumption charge   = $2114.10

Customer charge: = $ 115.00

                              Total charge before sales tax:
                              $1691.50 + $2114.10 + $115.00 = $3920.60

Sales tax:

                                        $3920.60(.06) = $ 235.24

*Total:**

                              $3920.60 + $235.24 = $4155.84

3.1.4 Large Industrial Rate Schedules

Large industry is a very unique customer for a utility. Often one or two large industries will purchase a significant portion of a utility's total generating capacity. Their
nature makes the billing more complex and important; consequently, it is absolutely
necessary to have a well-conceived and -designed rate schedule. It is important to
note that most utilities have very few customers that would qualify for or desire to be
on a large industrial rate schedule.

Typically, a large industrial schedule will include the same components as a
small industrial schedule. The difference occurs in the amount charged for each category. The customer charge, if there is one, tends to be higher. The minimum kw on
demand tends to be much higher, but all additional kw may be somewhat more economical (per kw) than on small industrial schedules. Similarly, the charge per kwh for
consumption can be somewhat less. The reason for this is economy of scale as it is
cheaper for a utility to deliver electrical energy to one large customer than the same
amount of energy to many smaller customers.

Figure 3–8 is an example of a large industrial schedule.

*Ignoring franchise payment and late charges.

Large Power and Light (LPL)

(TITLE AND/OR NUMBER)

Availability: Available on an annual basis by written contract to any retail customer. This schedule is not available for resale, standby, breakdown, auxiliary, or supplemental service. It is optional with the customer whether service will be supplied under this rate or any other rate for which he is eligible. Once a rate is selected, however, service will continue to be supplied under that rate for a period of 12 months unless a material and permanent change in the customer's load occurs.

Service will be supplied from an existing transmission facility operating at a standard transmission voltage of 69 kv or higher by means of not more than one transformation to a standard distribution voltage of not less than 2.4 kv. Such transformation may be owned by the company or customer. Service may be supplied by means of an existing primary distribution facility of at least 24 kv when such facilities have sufficient capacity.

Service will be furnished in accordance with the company's rules, regulations, and conditions of service and the rules and regulations of the Oklahoma Corporation Commission.

Net rate: Capacity charge:

$13,750.00: net per month for the first 2500 kilowatts (kw) or less of billing demand

4.20: net per month per kilowatt (kw) required in excess of 2500 kw of billing demand

.50: net per month for each reactive kilovoltampere (kvar) required above 60% of the billing demand

Plus an energy charge:

2.700¢: net per kilowatt-hour (kwh) for the first 1 million kwh used per month

2.570¢: net per kilowatt-hour (kwh) for all additional use per month

Determination of monthly billing demand: The monthly billing demand shall be the greater of (a) 2500 kw, (b) the monthly maximum kilowatt (kw) requirement, or (c) eighty percent (80%) of the highest monthly maximum kilowatt (kw) requirement established during the previous 11 billing months. The monthly maximum reactive kilovoltampere (kvar) required are based on 30-min integration periods as measured by appropriate demand indicating or recording meters.

Determination of minimum monthly bill: The minimum monthly bill shall consist of the capacity charge. The monthly minimum bill shall be adjusted according to adjustments to billing and kvar charges. If the customer's load is highly fluctuating to the extent that it causes interference with standard quality service to other loads, the minimum monthly bill will be increased $.50/kva of transformer capacity necessary to correct such interference.

Terms of payment: Payment is due within 10 days of the date of mailing the bill. The due date will be shown on all bills. A late payment charge will be assessed for bills not paid by the due date. The late payment charge shall be computed at $1\frac{1}{2}$% on the amount past due per billing period.

Adjustments to billing:

1. Fuel cost adjustment: The rate as stated above is based on an average cost of $2.00/million Btu for the cost of fuel burned at the company's thermal generating plants. The monthly bill as calculated under the above rate shall be increased or decreased for each kwh consumed by an amount computed in accordance with the following formula:

$$FA = A \times \frac{B}{10^6} \times C$$

where FA = fuel cost adjustment factor (expressed in dollars per kwh) to be applied per kwh consumed

A = weighted average Btu/kwh for net generation from the company's thermal plants during the second calendar month preceding the end of the billing period for which the kwh usage is billed

B = amount by which the average cost of fuel per million Btu during the second calendar month preceding the end of the billing period for which the kwh usage is billed exceeds or is less than $2.00/million Btu; any credits, refunds, or allowances on previously purchased fuel received by the company from any source shall be deducted from the cost of fuel before calculating B each month

C = ratio (expressed decimally) of the total net generation from all the company's thermal plants during the second calendar month preceding the end of the billing period for which the kwh usage is billed to the total net generation from all the company's plants including hydrogeneration owned by the company, or kwh produced by hydrogeneration and purchased by the company, during the same period

2. <u>Tax adjustment:</u> If there shall be imposed after the effective date of this rate schedule, by federal, state, or other governmental authority, any tax, other than income tax, payable by the company upon gross revenue, or upon the production, transmission, or sale of electric energy, a proportionate share of such additional tax or taxes shall be added to the monthly bills payable by the customer to reimburse the company for furnishing electric energy to the customer under this rate schedule. Reduction likewise shall be made in bills payable by the customer for any decrease in any such taxes.

Additionally, any occupation taxes, license taxes, franchise taxes, and operating permit fees required for engaging in business with any municipality, or for use of its streets and ways, in excess of two percent (2%) of gross revenues from utility business done within such municipality, shall be added to the billing of customers residing within such municipality when voted by the people at a regularly called franchise election. Such adjustment to billing shall be stated as a separate item on the customer's bill.

Figure 3-8   Large industrial rate schedule. (Courtesy of Oklahoma Gas and Electric Co.)

3.1.5 Cogeneration and Buy-Back Rates

Recently, there has been significant renewed interest in on-site-generated power. This can be from cogeneration (on-site generation of thermal heat and generation of electricity in some sort of cascading process), windmills, solar thermal, solar photovoltaics, or other sources. Generation of this energy for use only on site is often not cost effective due to variability of loads. Resale of excess electricity (when it is available) to the local utility, however, often makes the project economically feasible. To recognize this, utilities are developing buy-back rates for this excess electricity. Since the value of this energy may not really be the same as the cost to the utility of generating it, buy back usually requires a separate meter and a separate schedule.* Otherwise, the existing meter could simply be run backward.

At the time of this writing, these rates are so experimental and controversial that no samples will be given. Energy managers should realize, however, that these schedules do exist. A phone call to the local utility and/or state regulating board should yield copies.

3.1.6 Others

Many other schedules are being developed as the needs dictate. For example, one utility in the southwest has recently applied for a schedule affecting interruptible and curtailable loads. An *interruptible* load is one that can be turned off at certain times

---

*Remember, when the company generates an excess of electricity, it probably is not a peak time, so the utility really doesn't need it as badly.

of the day or year. This utility is willing to pay the company for not using that load during peaking times of the day or year.

A *curtailable* load is one that the company may be willing to turn off if given sufficient notice. For example, the utility may expect a severe peaking condition tomorrow. It may then call and ask that all the curtailable loads be turned off. Of course, the utility is willing to pay for this privilege.

In both cases, the utility pays for these loads by offering a reduction in the bill. In the case of curtailable loads, the reduction occurs every month during the peaking season whether or not the utility actually calls for the turnoff.

3.2 NATURAL GAS

Natural gas rate schedules are similar to electric rate schedules, but they tend to be much simpler. There is a peaking problem that is likely to occur on a very cold winter day and/or when supply disruptions exist. Due to the unpredictable nature of these problems, gas companies normally don't charge for peak demands. Instead, companies are placed into interruptible priority classes.

A customer with a high priority will not be curtailed unless absolutely necessary. A customer with the lowest priority, however, will be curtailed (normally some gas is supplied to keep pipes from freezing and pilot lights burning) whenever a shortage exists. To encourage use of the low-priority schedules, this gas is usually significantly cheaper. Most gas utilities have three or four priority levels. Some utilities allow customers to choose the appropriate schedule, while others strictly limit the choice. Figure 3–9 presents one example rate schedule for four priority levels. Here the customer has a limited choice, but there is some choice—priorities 3 and 4. Problem 3.6 demonstrates the use of this schedule.

Some points are demonstrated in this collection of schedules. First, the costs, in general, do decrease as the priority goes down. Of course, the probability of a curtailment dramatically increases. Second, the winter residential rate has an increasing block component on the top side. Only very large residential consumers would approach this block, so its intent is to discourage wanton utilization. Third, although no mention is made in Figure 3–9, fuel cost adjustments do exist, as do sales taxes. Again, some states do not or should not charge sales-tax on gas used directly in production.

3.3 FUEL OIL AND COAL

Fuel oils are a very popular fuel source in parts of the country, but they are rarely used in others. Basically, the availability and price of natural gas compared to the availability and price of fuel oil determine which is used.

Fuel oils can be classified as distillates or residuals referring to the refining or distillation process. Fuel oils 1 and 2 are distillates. No. 1 oil can be used as a domestic heating oil and diesel fuel. No. 2 oil is used by industry and in the home. The distil-

Residential priority 1	Commercial priority 2
Winter	Winter
First 1 ccf/month: $1.37	First 1 ccf/month: $1.37
Next 2.9 Mcf/month: $3.445/Mcf	Next 2.9 Mcf/month: $3.680/Mcf
Next 7.0 Mcf/month: $3.236/Mcf	Next 97/Mcf/month: $3.576/Mcf
All over 10 Mcf/month: $3.485/Mcf	Next 1900 Mcf/month: $3.353/Mcf
	All over 2000 Mcf/month: $3.266/Mcf
Summer	Summer
First 1 ccf/month: $1.37	First 1 ccf/month: $1.37
Next 2.9 Mcf/month: $3.445/Mcf	Next 2.9 Mcf/month: $3.68/Mcf
All over 3.0 Mcf/month: $3.393/Mcf	Next 97 Mcf/month: $3.576/Mcf
	Next 100 Mcf/month: $3.353/Mcf
	All over 200 Mcf/month: $3.185/Mcf
Industrial priority 3	Industrial priority 4
First 1 ccf/month: $1.37	First 4000 Mcf/month or fraction thereof: $11,712
Next 2.9 Mcf/month: $3.65/Mcf	All over 4000 Mcf/month: $2.928/Mcf
Next 97 Mcf/month: $3.321/Mcf	
All over 100 Mcf/month: $3.049/Mcf	

Figure 3–9 Gas schedules for one utility. (Courtesy of Oklahoma Natural Gas Co.)

lates are easier to handle and require no heat to maintain a low viscosity (i.e., they can be pumped or poured with ease).

Residual fuel oils include nos. 4, 5, and 6. Optimum combustion is more difficult to maintain in these oils due to variation in their characteristics that result from different crudes and refining processes. No. 6 or residual bunker C is a very heavy residue left after the other oils have been refined. It has a very high viscosity and must be heated in cold environments to maintain a *pour point* (usually somewhere around 55°F).

The sulfur content of fuel oil normally ranges from .3 to 3.0%. Distillates have lower percentages than residuals unless the crude has a very high sulfur level. Sulfur content can be very important in meeting environmental standards and thus should be watched carefully.

Billing schedules for fuel oils vary widely among geographical areas of the country. The prices are set by market conditions (supply vs. demand), but within any geographical area they are fairly consistent. Within each fuel oil, there is a large number of sulfur grades, so shopping around can sometimes pay off. Basically, the price is simply a flat charge per gallon, so the total cost is the number of gallons used times the price per gallon.

Coal is very similar to fuel oil in that it is plagued with varying grades and varying sulfur content. It is, in general, less expensive than fuel oil, but it does require higher capital investments for pollution control, coal receiving and handling equipment, storage, and preparation. Transportation costs will also likely be higher.

Finally, coal is likely to have a significantly higher loss due to unburned combustibles. If combustion air is controlled, natural gas has almost none, while fuel oil has only a small amount. Coal, however, is much more difficult to fully combust. Coal is priced on a per ton basis with provisions for or consideration of sulfur content and percent moisture.

3.4 STEAM AND CHILLED WATER

In some areas of the country, it is possible to purchase steam and chilled water directly instead of buying the fuel and generating your own. This can occur in the case of large-scale cogeneration (steam), refuse-fueled plants (steam), or simple economics of scale (steam and/or chilled water). In both cases (steam and chilled water), it is normal to charge for the energy itself (pounds of steam or ton-hours of chilled water) and the demand (pounds of steam per hour or tons of chilled water demand). A sample hypothetical rate schedule is shown in Figure 3–10.

Obviously, these rates are competitive with self-generated rates and purchase of

Steam

Steam demand charge:
$1500.00/month for the first 2000 lb/h of demand or any portion thereof
$550.00/month/1000 lb/h for the next 8000 lb/h of demand
$475.00/month/1000 lb/h for all over 10,000 lb/h of demand

Steam consumption charge:
$3.50/1000 lb for the first 100,000 lb of steam per month
$3.00/1000 lb for the next 400,000 lb of steam per month
$2.75/1000 lb for the next 500,000 lb of steam per month
$2.00/1000 lb for the next 1 million lb of steam per month

Negotiable for all over 2 million lb of steam per month

Chilled water

Chilled water demand charge:
$2500.00/month for the first 100 tons of demand or any portion thereof
$15.00/month/ton for the next 400 tons of demand
$12.00/month/ton for the next 500 tons of demand
$10.00/month/ton for the next 500 tons of demand
$9.00/month/ton for all over 1500 tons of demand

(One ton is defined as 12,000 Btu/h, and an hour is defined as any 60 consecutive min.)

Chilled water consumption charge:

$.069/ton · h for the first 10,000 ton · h/month

$.06/ton · h for the next 40,000 ton · h/month

$.055/ton · h for the next 50,000 ton · h/month

$.053/ton · h for the next 100,000 ton · h/month

$.051/ton · h for the next 100,000 ton · h/month

$.049/ton · h for the next 200,000 ton · h/month

$.046/ton · h for the next 500,000 ton · h/month

Base rates: Consumption rates subject only to escalation of charges listed in conditions of service and customer instructions

Figure 3–10 Hypothetical steam and chilled water rate schedule.

steam. Chilled water conserves considerable amounts of capital and maintenance monies. In general, when steam-chilled water is available, it is worthy of consideration. The primary disadvantage is that the user doesn't have control of the generating unit, but that has always been true of electricity and natural gas.

3.5 SUMMARY

In this chapter we have gone into considerable depth in analyzing rate schedules and costs for electricity, gas, coal, fuel oil, steam, and chilled water. A complete understanding of all the rate schedules is vital for an active and successful energy management program.

In the past, few managers have even understood the various components of these rate schedules, and even fewer have ever seen their own rate schedules. The future successful manager will not only be familiar with the terms and the schedules themselves, but he or she will also likely work with utilities and rate commissions toward fair rate-setting policies.

QUESTIONS

3.1. Recently, there has been a trend across the country for utilities to charge more for demand and keep consumption billing about the same (or even reduce the charges). Discuss why this may be occurring.

3.2. Discuss why demand control during peaking months may be more profitable than during nonpeaking months. How might a ratchet clause affect this?

3.3. Discuss ways a manufacturing company might prepare for natural gas curtailments to minimize their impacts.

3.4. Discuss why some managers have failed to analyze and understand their energy rate schedules.

3.5. Do you think a company should periodically analyze their energy rate schedules to see if a change is in order? Explain.

3.6. Discuss why cogeneration, wind, and solar buy-back rates are not as high as what the utility charges customers.

PROBLEMS

3.1. In working with Ajax Manufacturing Company, you find six large exhaust fans exhausting general plant air (not localized heavy pollution). They are each powered by 30-hp electric motors. You find they can be turned off periodically with no adverse effects. You place them on a central timer such that each one is turned off for 10 min each hour. At any time, one of the fans is off, and the other five are running. The fans operate 10 h/day, 250 days/year. Assuming the company is on the rate schedule given in Figure 3–6, what is the total dollar saving per year to the company? The company is on service level 3 (distribution service) and is well above the 75-kw minimum demand. Neglect any ratchet clauses. (There will be significant heating savings since conditioned air is being exhausted, but ignore these for now.)

3.2. A large manufacturing company in southern Arizona is on the rate schedule shown in Figure 3–6 (service level 5, secondary service). Their peak demand history for last year is shown below. They have found a way to reduce their demand in the off-peak season by 100 kw, but the peak season demand will be the same (i.e., the demand in each month of November through May would be reduced by 100 kw). Assuming they are on the 65% ratchet clause specified in Figure 3–6, what is their dollar savings? Assume the high month was July of the previous year at 1150 kw. If the demand reduction of 100 kw occurred in the peak season, what would be the dollar savings (i.e., the demand in June through October would be reduced by 100 kw)?

Month	Demand (kw)	Month	Demand
Jan.	495	July	1100
Feb.	550	Aug.	1000
March	580	Sept.	900
April	600	Oct.	600
May	610	Nov.	500
June	900	Dec.	515

3.3. In the data for Problem 3.2, how many months would be ratcheted, and how much would the ratchet cost the company above normal billing?

3.4. In working with a company, you find they have averaged 65% power factor over the past year. They are on the rate schedule shown in Figure 3–6 and have averaged 1000 kw/month. Neglecting any ratchet clause and assuming their demand and power factor is constant each month, calculate the savings for correcting to 80% power factor. How much capacitance would be necessary to obtain this correction? Assume they are on transmission service, PLT (level 1).

3.5. A company has contacted you regarding their rate schedule. They are on the rate schedule shown in Figure 3–6, service level 5 (secondary service), but are near transmission lines and so can accept service at a higher level (service level 1) if they buy their own transformers. Assuming they consume 300,000 kwh/month and are billed for 1000 kw each month, how much could they save by owning their own transformers. Ignore any charges other than demand and energy.

3.6. In working with a brick manufacturer, you find for gas billing that they were placed on an industrial (priority 3) schedule (see Figure 3–9) some time ago. Business and inventories are such that they could switch to a priority 4 schedule without many problems. What is the savings? They consume 7000 Mcf of gas per month for process needs and essentially none for heating.

Chapter 4
....

Economic Evaluation

4.0 INTRODUCTION

Most expenses undertaken in the name of energy management must be justified in terms of avoided costs. Usually the expenses come at the beginning of a project with the benefits coming later, and the question arises of how to add the cash flows in light of the different times of their occurrence. Clearly there is a problem—a sum of money at the beginning of the year buys less at the end of the year and is worth even less at the end of 5 years—so adding cash flow amounts at the start of a project to cash flow amounts several years later is adding unlike quantities. But both quantities are cash, and there should be some way to compare them. The method developed to analyze this kind of problem has been given the names *engineering economy* and *discounted cash flow analysis*. In this chapter we apply discounted cash flow principles to problems of energy management with particular emphasis upon inflation and after-tax analysis. In Section 4.1 we give definitions of costs encountered in energy management studies and present some of the basic sources for estimates of these costs. In Section 4.2 we present some of the classical ideas of engineering economy and include sections on cash flow analysis, comparing alternative patterns of cash flow and determining rates of return. In Section 4.3 we show how to take taxes into account. In Section 4.4 we show how to incorporate inflation into economic analyses and discuss the difference between current dollars and constant dollars as they apply to energy management analyses.

Throughout the chapter, formulas are given where appropriate. The authors have avoided the use of the customary tables because most of the users of this book have access to either small financial calculators or a microcomputer with financial programs and because the use of tables often obscures the pattern of cash flows to the detriment of financial planning.

In any discipline there is an underlying methodology. The guiding steps of discounted cash flow analysis are the following:

Step 1: Define the Alternatives.   The problem is stated and the list of possible solutions is refined to those worthy of analysis.

Step 2: Estimate Relevant Costs.   Each alternative is defined in terms of its cash flows including the amount, timing, and direction (benefit or cost) of each.

Step 3: Analyze Viable Alternatives.   The cash flows identified in step 2 are compared and combined using the discounted cash flow methods described in this chapter. Noneconomic factors are brought into the analysis in as much detail as necessary.

Step 4: Perform Sensitivity Analyses.   The cash flow analysis is examined in order to determine the effects of uncertainties in the estimates.

The tools needed to perform these steps are now presented.

4.1 COSTS AND COST ESTIMATION

The first step in a discounted cash flow analysis is to choose alternatives for analysis. To a great extent that step is the subject of this entire book. The second step is the systematic study of the costs of each alternative. This topic is the subject of this section.

Costs can be categorized in several ways, depending on the purpose of the analysis. For the purpose of comparing energy projects, we choose to divide them into costs at the start of a project, costs that take place throughout the duration of the project, and salvage costs.

4.1.1 Initial Costs

Initial costs are defined to be the costs that occur at the beginning of a project. These costs include purchase and installation, minus any salvage value obtained by selling the item for which the new one is a replacement. Any tax credits on the initial investment are also subtracted from this initial cost (see Section 4.3). These costs also include any engineering that must be done, any permits that must be obtained, and any renovation that must be performed before a project can be started. Methods of estimating these costs range from approximations based on past experience (rules of thumb) to detailed analysis in which as many factors as possible are taken into account. (See Reference 1 for a good treatment of this subject.) The accuracy of the estimate, and the cost of the effort used to derive the estimate, should depend on the consequences of making an error based on the estimate. If the estimate is used to determine whether the project is feasible, approximate methods are good enough. If the estimate is used to obtain financing for the project, the estimates should be the best possible.

For analysis purposes, capital costs are assumed to take place at the end of year 0 within the project life.

4.1.2 Annual Costs or Benefits

In addition to capital costs, other costs occur on a regular basis. Examples of such costs are operating and maintenance costs. These include the costs of labor and material that can be directly attributed to a project (*direct costs*) and costs that the project shares with other projects such as supervision, janitorial supplies, and the like— so-called *indirect costs*. Energy costs are regarded as indirect costs in companies where the cost of the energy needed to produce one unit of output is not known; where the energy cost per unit of product is known (such as in steel or aluminum plants), these costs are often regarded as direct costs.

It is difficult to estimate annual costs accurately. When the costs involve building-related repair and maintenance, the *Unit Price Standards Handbook* of the U.S. Department of Defense is excellent [2]. This book gives labor hour estimates by craft and specifies materials and special equipment needed for each task. Another source for this information is Sack [3]. Also, vendors usually have estimates of the time and skill levels required to maintain their products, although they cannot always be relied upon.

Annual costs are usually treated as occurring at the end of the year in which they are incurred.

4.1.3 Salvage Values

Most equipment items have cash values at any given time after their purchase, and this *salvage value* must be taken into account in any cash flow analysis. Unfortunately, the salvage value depends on many unknowns—the market for used equipment, how the equipment was used, the equipment maintenance, and more. If the salvage value is a deciding factor in the comparison of two alternatives (and it usually is *not*), it can be varied in the analysis to see whether to spend additional time and money to better define it. Trade magazines carry advertisements for used equipment, and they can be used to estimate the salvage value.

A salvage value is treated as occurring at the end of the year in which the item is sold.

Note: Under the present tax system (1984), salvage values are considered as zero when the tax consequences of depreciation are taken into account. If an equipment item is sold after it has been fully depreciated and the price is less than the original purchase price, the revenue from the sale is considered as ordinary revenue in the year the sale occurs and is taxed accordingly. In other circumstances, see a good accountant or a text on engineering economy such as Reference 4.

4.1.4 Energy Costs

Energy costs depend on the type of energy (gas, electricity, coal, steam, etc.) and on the billing schedule of the particular utility for the particular energy form in use. For gas, steam, liquified natural gas (LNG), and many other fuels, the bill is dependent on the total amount of energy consumed. Usually, the first units consumed cost the most per unit with additional units of energy costing less. As discussed in Chapter 3, the electrical bill includes two components: the *energy* charge, based on the total amount of energy consumed during the month and in some way proportional to the amount of fuel consumed to generate the electricity, and the *demand* or *power* charge, based on the peak demand during the previous month (or year) and reflecting the cost of providing the generating capacity needed to meet the peak demand. Another relevant number that often enters electrical billing is the *power factor*—the ratio between the resistive power measured by the meter and the total power delivered (the vector sum of the resistive and inductive power). If the power factor is too low— say, less than 85%—the utility adds a specified amount onto the bill. These and other relevant factors are made specific in the *billing schedule,* a legal document with the exact formula used to calculate electric bills. There are normally many such schedules, each with slightly different prices, and an examination of relevant billing schedules may lead to a reduction in cost. Collecting and examining the rate schedule is thus an important step in an energy analysis.

To see how an electric bill is calculated, consider the calculation of an electric bill for a customer with a January demand of 140,000 kwh, a peak 15-min demand during January of 500 kw, and a power factor of 80%, under the electrical schedule of Figure 4-1 (typical of many such schedules). The bill is calculated in Figure 4-2.

Energy charge
_____

$.06/kwh for first 500 kwh (block 1)

$.12/kwh for next 4500 kwh (block 2)

$.10/kwh for next 10,000 kwh (block 3)

$.07/kwh for all remaining kwh (block 4)

Demand charge
_____

$12.00/kw for first 450 kw of billed demand in excess of 5 kw (block D1)

$9.00/kw for all remaining kw of billed demand (block D2)

Billed demand and power factor correction
_____

The billed demand is based on the peak 15-min electrical demand during the previous month as computed by the company's meters, increased by $\frac{3}{4}$% for each percent by which the power factor is less than 90%.

Figure 4-1 Typical billing schedule for use in calculating a monthly electric bill.

Billed demand = 500 × [1 + .0075 × (90 − 80)]
= 538 kw

Note: Although the demand for any measure could legitimately be carried out to several significant figures, the convention to be used here will be to round any demand to the nearest integer.

Energy charge:

Block	kwh in block	Cost/kwh	Cost of block	kwh remaining
1	500	$.06	$30.00	139,500
2	4,500	$.12	$540.00	135,000
3	10,000	$.10	$1,000.00	125,000
4	125,000	$.07	$8,750.00	0
		Total energy charge:	$10,320.00	

Demand charge:

Block	kw in block	Cost/kw	Cost of block	kw remaining
D1	445 (= 450 − 5)	$12.00	$5,340.00	93
D2	88	$9.00	$792.00	0
		Total demand charge:	$6,132.00	

Total monthly bill (excluding taxes) = $16,452.00

Figure 4-2 Calculation of a typical monthly electric bill.

The billing schedule can be used to estimate energy cost savings associated with a particular energy management measure. The specific benefits are first listed. Next the appropriate utility schedules are consulted to determine the amount of annual savings in terms of the present rates. Third, an inflation estimate is obtained for each energy source and is used to estimate the energy cost saving for each year out to the planning horizon. These savings are then incorporated into the economic analysis.

One important note: The rates used in calculating the energy cost savings should reflect the costs actually avoided; i.e., the decrease in the projected costs should be computed at the lowest kwh cost rate paid by the company, rather than some weighted average of the higher-cost energy cost blocks. For example, if the usual monthly energy bill is based on 400,000 kwh and a 1000 kw peak demand, and a load of 10 kw is removed for 40 h each week, 4.33 weeks/month, then 1732 kwh are removed from the energy charge. Note that these affect the bill only at the *lowest* cost per unit, i.e., in the preeceding example, at 7.0 cents/kwh. Note also that the 10 kw can be deducted from the demand only if it is known for sure that removing this particular load of 10 kw will reduce the peak on which the demand charge is calculated. These may be obvious observations, but in our experience, companies have found them easy to neglect.

As with other annual expenses, energy expenses or savings are usually treated as occurring at the end of each year.

4.2 ECONOMIC ANALYSIS, WITHOUT INFLATION

In Section 4.1, costs and cost estimation were discussed as they relate to energy management. Particular attention was paid to the most obvious way in which energy cost savings are realized—through reductions in utility costs. In this section, basic concepts of discounted cash flow analysis are introduced. Taxes and inflation are left to Section 4.3.

4.2.1 Data Collection and Presentation

Once cost data have been collected, the next step is to prepare the data in a form useful for analysis. All the cash flows in a given year must be labeled as to type, amount, and direction (revenue or cost). A typical example is given in Figure 4–3. This method has the advantage that cash flows arranged in this way can be made into equivalent cash flows (see below) with the use of very simple formulas and without the use of tables.

Year	First cost	Operating and maintenance	Energy cost savings	Net cash flow
0	−$150,000			−$150,000
1		−$10,000	$45,000	$35,000
2		−$11,000	$50,000	$39,000
3		−$12,000	$56,000	$44,000
4		−$13,000	$62,000	$49,000
5	$30,000 (salvage value)	−$14,500	$69,000	$84,500

Figure 4–3 Example of net cash flows.

4.2.2 Equivalence of Cash Flows

The original challenge of this chapter was to compare two cash flows occurring at different times. This challenge can be met through the use of a *discount rate.* One example of a discount rate is basic bank interest. If, for example, a person can get 10% annual interest on his or her money at a bank, then $1000 now is worth $1100 1 year from now, and the two amounts can be considered *equivalent* at 10%. If, in another example, the annual inflation rate is 12%, then $1000 now has the same purchasing power as $1120 1 year from now, and these two amounts can be considered equivalent at 12%. In general, an amount P now is equivalent to an amount F, n periods in the future, at a discount rate 100 I% per period if the following relationship holds:

$$F = P(1 + I)^n \qquad\qquad (4\text{--}1)$$

End of year	Value
0	$1000
1	$1000 \times (1 + .12) = \$1120$
2	$1120 \times (1 + .12) = 1000 \times (1 + .12)^2$ = 1254
3	$1254 \times (1 + .12) = 1000 \times (1 + .12)^3$
...	....
n	$1000 \times (1 + .12)^n$

Figure 4-4 Growth of compound interest.

Another way of stating this is to define the *equivalent present worth* of a cash flow *F*, occurring *n* years in the future, by Equation 4-2, where *I* is the annual discount rate:

$$P = \frac{F}{(1 + I)^n} \qquad (4\text{-}2)$$

Equation 4-2 is reasonable in the many situations where the return on an investment can be considered analogous to the return on a savings account in a bank, with the interest left in the bank. Suppose that $1000 is invested now, at an annual interest rate of 12%. This investment will grow as shown in Figure 4-4. Note that this growth can be expressed in terms of equation 4-2. Not all investments give a return as easily identified as bank interest, but the value of the analogy is overwhelming.

[*Note:* If the return is compounded monthly with a nominal annual rate *I*, then the equivalent annual rate is $(1 + I/12)^{12}$. Under usual circumstances, annual compounding—using *I* as the return rate per year and *n* as the number of years—is adequate for computational purposes.]

Equation 4-2 can be used to determine the equivalent present worth of cash flows shown in Figure 4-3, as shown in Figure 4-5. A discount rate of 12% was used. Since the total equivalent present worth is positive at a 12% discount rate, the original investment of $150,000 yields at least a 12% *rate of return*, before inflation. To find the actual rate of return of this investment, replace the 12% by successively larger values until the total present worth is zero. For the preceding investment, this rate of return is 17.28%.

Year	Net cash flow	Discount factor	Equivalent present worth
0	−$150,000	$1/(1 + .12)^0$	−$150,000
1	35,000	$1/(1 + .12)^1$	31,250
2	39,000	$1/(1 + .12)^2$	31,090
3	44,000	$1/(1 + .12)^3$	31,318
4	49,000	$1/(1 + .12)^4$	31,140
5	84,500	$1/(1 + .12)^5$	47,948
		Total:	$22,746

Figure 4-5 Calculating the equivalent present worth of a cash flow stream.

Present worth equivalence is an important form of economic equivalence. Another important form is *annual worth* equivalence. The equivalent uniform annual worth of a cash flow stream is the one annual value, taken at the end of every year of the project, that would create cash flows equivalent in present worth to the original cash flow stream. Examples of such equivalent uniform payments are mortgage payments, where the payments cover the initial price of a house as well as annual payments of taxes and insurance, and car lease agreements, where the lease payments cover the purchase price of the car minus the estimated value of the car at the time it is returned as well as a profit for the lessor.

Equation 4-3 relates a present worth P to its equivalent uniform annual worth A [4]:

$$A = P \times \frac{I \times (1 + I)^n}{(1 + I)^n - 1} \qquad (4\text{-}3)$$

where I = discount rate per period
n = number of periods

Equation 4-3 can be used to determine the equivalent annual worth of the cash flow stream of Figure 4-3 in either of two ways: (1) The individual cash flows can be converted into their equivalent present worth cash flows using Equation 4-2, then into equivalent annual worth cash flows using Equation 4-3, and, finally, summed, or (2) Equation 4-3 can be used directly on the equivalent present worth of Figure 4-5. Using the latter procedure gives (assuming a 12% discount rate)

$$A = 22{,}746 \times \frac{.12 \times 1.12^5}{1.12^5 - 1} \qquad (4\text{-}4)$$
$$= 6310$$

Either the present worth method or the equivalent uniform annual worth method may be used when comparing projects or equipment items with the same service lives. When comparing projects that seem to have different service lives, define what is to be done between the end of the shorter project and the end of the longer project and add that cost to the cost of the shorter project. This method can be cumbersome, but it is the only easy way to deal with costs or benefits that are inflating at different rates throughout the length of the project.

If the projects are both serving a single function that will not be renewed or if it is known that the equipment will not be replaced at the end of its service life, then the service life for comparison is taken as the longer of the two service lives with all cash flows after the end of the shorter service life assumed to be zero. By using this longer service life, either an annual worth or a present worth comparison can be used. For a more detailed explanation of this problem, see Reference 5.

4.2.3 Calculating Equivalent Present Worth for a Project

Before proceeding further, consider the way these calculations could be applied to a particular project. This project costs $700,000, has a salvage value at the end of year 5 estimated as $100,000, will incur a maintenance cost of $60,000 each year beginning

Year	One-time costs	Maintenance	Return from project	Net cash flow (NCF)	Discount factors	Discounted cash flow
0	−700	—	—	−700	1	−700.0
1	—	—	—	0	1/1.15	0
2	—	−60	—	−60	$1/1.15^2$	−45.4
3	—	−60	450	+390	$1/1.15^3$	+256.4
4	—	−60	540	+480	$1/1.15^4$	+274.4
5	100	−60	648	+688	$1/1.15^5$	+342.1

Net present worth (× 1000): $127.5

Figure 4-6   Present worth calculations without taxes or inflation.

in year 2, and will return savings starting at $450,000 during year 3 and increasing 20% each year thereafter. For purposes of the problem, assume that the discount rate used to calculate the present worth is 15% and that inflation is neglected. The cash flows of this project can be arranged and the calculations performed as shown in Figure 4-6. Since the net present worth of this project is positive at a 15% discount rate, the rate of return of this project is greater than 15%. If a company considers projects only with a rate of return of at least 15%, this project should be taken into consideration. It might be worthwhile, moreover, to recalculate these figures at higher rates of return to find out exactly how attractive this investment is.

4.3 TAXES

Most companies are taxed, by local, state, and federal governments. This taxation varies in what is taxed, what tax credits are given, and when tax amounts and credits occur. The variation is by location, by governmental unit, and by year, but the principles remain the same. In general, a corporate tax rate is applied only to profit, so anything that decreases profit decreases the tax amount and thus creates a tax credit. In general, capital investments with a life of 5 years or longer can be depreciated, and the depreciation can be treated as an operating expense and thus can create a tax credit. Often there are also special tax credits designed to increase the desirability of a particular kind of investment—an investment tax credit of 10% of the original purchase price and credits of 50% of the purchase price of solar heating units are examples—and the details change from year to year and place to place. Before performing an after-tax economic analysis, it is essential to determine (from an accountant or a tax lawyer) what tax credits are applicable to the particular situation under study. Then these taxes must be taken into account.

Consider the example presented earlier, with the cash flows shown in Figure 4-6. Suppose that the corporate tax rate is 48%, that any capital expense with a life of 5 years or more is eligible for a tax deduction of 10% of the purchase price the first year, and that the minimum attractive after-tax rate of return is 10%. Any cash flow that increases revenue also increases taxes, in an amount equal to the corporate tax rate multiplied by the savings. Any cash flow that increases costs decreases taxes, in an amount equal to the corporate tax rate multiplied by the increase in costs. Thus to calculate the cash flow of the tax, multiply any before-tax cash flow that increases costs or revenues by the corporate tax rate and change the sign.

In addition to these cash flows, there is depreciation. Under the Tax Equity and Fiscal Responsibility Act of 1982, equipment is divided into four tax life categories: 3-year, 5-year, 10-year, and 15-year. Most machine tools and many energy management capital expenses have a 3-year tax life; most other capital expenses are in the 5-year tax category. (For more detail, see a good tax accountant.) The depreciation percentages each year for each category are given in Table 4-1. In calculations, these percentages are applied to the original capital expense *without regard to the salvage value.* Furthermore, the tax life category for the equipment is that specified by the law; the actual life anticipated is irrelevant. The cash consequences of this depreciation schedule are credits equal to the corporate tax rate multiplied by the depreciation for the given year. (Although depreciation is not a cash flow, it is regarded by the government as an expense; it therefore lowers the taxable profit and creates a tax credit.)

In the example, suppose (1) that the project consists of equipment all of which falls into the 5-year category, (2) that the company is subject to taxes which total 48%, and (3) that these taxes are all based on the net profit. Then the cash flows associated with the $700,000 investment are given by $700,000 \times .48 \times P_i$, where P_i is the depreciation percentage for year i. Here $P_1 = .15$, $P_2 = .22$, etc. These tax credits are $50,400, $73,920, $70,560, $70,560, and $70,560 for years 1–5, respectively.

The general procedure has the following steps:

1. Estimate the before-tax cash flows for each year of the project life, and list the net cash flow for each year.

TABLE 4.1 DEPRECIATION PERCENTAGES (ACCELERATED CASH RECOVERY SCHEDULE) UNDER THE TAX EQUITY AND FISCAL RESPONSIBILITY ACT OF 1982

Recovery year	Property class			15-year public utility
	3-year	5-year	10-year	
1	25	15	8	5
2	38	22	14	10
3	37	21	12	9
4	—	21	10	8
5	—	21	10	7
6	—	—	10	7
7	—	—	9	6
8	—	—	9	6
9	—	—	9	6
10	—	—	9	6
11	—	—	—	6
12	—	—	—	6
13	—	—	—	6
14	—	—	—	6
15	—	—	—	6

Year	One-time costs	NCF* before taxes	Taxes	Cash flow from deprec.	NCF after taxes	Discount factors	Discounted after-tax cash flow
0	−700	−700	+70.0	−	−630.0	1.0000	−630.0
1	−	0	0	50.4	+50.4	.8696	+43.8
2	−	−60	+28.8	73.9	+42.7	.7561	+32.3
3	−	+390	−187.2	70.6	+273.4	.6575	+179.8
4	−	+480	−230.4	70.6	+320.2	.5718	+183.1
5	100	+688	−330.2	70.6	+428.4	.4972	+213.0

Present worth (× 1000): $22.0

*Net cash flow

Figure 4-7   Present worth calculations after taxes, without inflation.

2. Calculate the tax savings or cost associated with each cash flow, and list the net amount of tax consequence (positive or negative) for each year.
3. Add the results from step 1 and step 2, and list this amount for each year. This is the net after-tax cash flow.
4. For each year, find the discount factor necessary to equate cash flows for that year to present value cash flows, at a prescribed rate of return. The formula was presented earlier:

$$P = \frac{F}{(1 + i)^n}$$

where P is the present worth equivalent, F the net after-tax cash flow, i the rate of return, and n the number of years the payment is to be discounted.
5. Multiply the net after-tax cash flows by the appropriate discount factors, and add the results.

This procedure is illustrated in Figure 4-7 for the example presented.

4.4 INFLATION

Inflation should be taken into account in most economic analyses. Otherwise, all cash flows are treated as if they were inflating at the same rate, and the results of the analysis may be distorted. In this section, the procedures of the previous sections are modified to take inflation into account.

4.4.1 Terminology

The only new definitions needed are those of *current dollars* and *constant dollars*. *Current dollars* are the dollars that are actually spent at a given time. These dollars have a purchasing power that decreases with inflation. In contrast, *constant dollars* are dollars with a constant purchasing power. These dollars represent purchasing power at a particular time and are generally labeled by that time, for example, *1980 dollars*. In the days of fixed-rate mortgages, it was possible to buy a house for a specific amount and to pay off the mortgage at a constant amount per month, thus paying off a loan in an amount that decreased in value each month. Thus the mortgage holders were paying back an amount originally calculated in constant dollars and were doing so in current dollars. So long as the lenders could borrow money at lower rates than the money they were lending, this system was viable. As soon as the mortgage rates dropped below the minimum rate at which the lenders could borrow money, however, the lending institutions were in trouble. This kind of problem led to some of the major financial reforms of the early 1980s.

The relationship between a current dollar amount D_{curr} and an equivalent constant dollar amount D_{cons} is given by

$$D_{cons} = \frac{D_{curr}}{(1 + I)^n} \tag{4-5}$$

where n is the number of years between the year when the money is to be spent and the reference year for D_{cons} and where I is the applicable inflation rate. Note that Equation 4-5 is equivalent to Equation 4-2 except that the rate of compounding i is replaced by the inflation rate I.

4.4.2 Incorporating Inflation into Economic Studies

An extension of the method previously described provides insight into the computations involved and is easy to check. This extension to problems involving inflation proceeds as follows:

Step 1.   Calculate the current-dollar before-tax cash flows as before, this time taking into account any predicted inflationary effects. In particular, care should be taken to obtain some estimate for inflation rates of utilities, labor, and equipment and to include these rates explicitly in the estimation of current-dollar costs.

Step 2.   For each year, add the cash flows to obtain net current-dollar cash flows before taxes.

Step 3.   Apply relevant tax factors and depreciation to determine after-tax cash flows, still in current dollars.

Step 4.   Use Equation 4-5, with the projected inflation rate in the consumer price index I, in order to calculate discount factors for each year. These factors, when multiplied by current-dollar figures, can yield their constant-dollar equivalents. If the company doing the analysis uses a current-dollar rate of return, however, this percentage should be used as I, and the sum of the cash flows discounted by I will be the present worth equivalent of this cash flow. In that case, step 5 is not needed.

Step 5.   Calculate for each year the discount factors corresponding to the chosen constant-dollar rate of return. For each year, multiply the discount factor calculated in step 4 by the new discount factor; the result is a set of factors that yield the discounted constant-dollar present worth for any after-tax cash flow.

Step 6.   Using the after-tax net cash flows and the discount factors from step 5, compute the discounted after-tax constant-dollar net cash flow.

After all these computations, the basic form is still that of Figure 4-7, with different values for the NCF before taxes and with different discount factors.

As an example of the application of these steps, consider the following situation: A proposed energy management measure can be initiated now with full operation expected to start 1 year from now. At present prices, the electrical savings will be $30,000/year and the labor cost will be $10,000/year. The equipment will cost $50,000 and is in the 3-year tax life category. The equipment is expected to last 5 years. The present salvage value of comparable 5-year-old equipment is $10,000 and is increasing at 6%/year. The planning horizon is 5 years. Projected inflation rates are as follows: electricity, 20%; labor, 4%; general inflation rate, 4%. The applicable tax rate is 48%, the prescribed after-tax constant-dollar rate of return is 15%, and a tax credit of 10% of the initial cost is available the year the equipment is purchased. Is the investment worthwhile? The calculations are shown in Figures 4-8, 4-9, and 4-10. Since the investment shows a positive present worth at the after-tax constant-dollar rate of return of 15%, the investment is yielding at least that great a rate of return.

Year	Electrical savings	Labor costs	NCF
1	0	0	0
2	$36,000 (= 30,000 × 1.2)	−$10,400 (= 10,000 × 1.04)	$25,600
3	$43,200 (= 30,000 × 1.2^2)	−$10,816 (= 10,000 × 1.04^2)	$32,384
4	$51,840	−$11,249	$40,591
5	$62,208	−$11,699	$50,509

Figure 4-8   Calculation of current-dollar net cash flow (excluding one-time costs).

Year	Current to constant dollars	Constant to present worth dollars	Summary discount rate
0	1.0000 = 1/1.04^0	1.0000 = 1/1.15^0	1.000
1	.9615 = 1/1.04^1	.8696 = 1/1.15^1	.8361
2	.9246 = 1/1.04^2	.7561 = 1/1.15^2	.6991
3	.8890	.6575	.5845
4	.8548	.5718	.4887
5	.8219	.4972	.4086

Figure 4-9   Calculation of summary discount rate.

Year	One-time costs	NCF before taxes	Taxes	Cash flow from deprec.	NCF after taxes	Discount factors	Discounted after-tax cash flow
0	−50.0	−50.0	+5.0	−	−45.0	1.0000	−45.0
1	−	0	0	+6.0	+6.0	.8361	+5.0
2	−	+25.6	−12.3	+9.1	+22.4	.6991	+15.7
3	−	+32.4	−15.6	+8.9	+25.7	.5845	+15.0
4	−	+40.6	−19.5	−	+21.1	.4887	+10.3
5	13.4 ($= 10 \times 1.06^5$)	+63.9	−30.7	−	+33.2	.4086	+13.6

Present worth ($\times$ 1000): $14.6

Figure 4-10   Present worth calculations after taxes, with inflation.

4.5 SUMMARY

This chapter has given the economic analysis method needed in order to evaluate projects which include cash flow components inflating at different rates. Taxes have also been included, as they must be in most economic analyses. The method is easy to understand and will be used throughout the rest of this book.

REFERENCES

1. Phillip F. Ostwald, *Cost Estimating for Engineering and Management,* Prentice-Hall, Englewood Cliffs, N.J., 1974.
2. *Unit Price Standards Handbook,* Stock No. 008-047-00221-1, Washington, D.C., Nov. 1977 (also known as NAVFAC P-716, ARMY TB 420-33, and AIR FORCE 85-56).
3. Thomas F. Sack, *A Complete Guide to Building and Plant Maintenance,* Prentice-Hall, Englewood Cliffs, N.J., 1971.
4. John A. White, Marvin H. Agee, and Kenneth E. Case, *Principles of Engineering Economic Analysis,* Wiley, New York, 1977.
5. Donald G. Newnan, *Engineering Economic Analysis,* Engineering Press, San Jose, Calif. 1976.

QUESTIONS

4.1. The opening sentence refers to *avoided* costs. Why is this term more correct than reduced costs?

4.2. Some examples of gas and electrical billing use a so-called "declining block" utility rate, i.e, a rate where the cost per unit declines as more is consumed. Give substantial arguments for and against this type of rate, and then decide whether *you* favor it.

4.3. Why would a company use a different acceptable rate of return for energy management projects than for other projects? If you don't understand the answer to this question, then you will have a difficult time defending your projects against these arguments.

4.4. Which are more important in the budget decisions for a state, economic criteria or non-economic criteria? Under what circumstances does one group of criteria predominate?

4.5. Why should only the equivalent uniform annual cost method be used in comparing projects of unequal service lives?

4.6. What are some good sources for inflation rate projections?

PROBLEMS

4.1. Given the following schedule for gas, calculate the monthly gas bill for a monthly consumption of 8500 Mcf:

First	1 ccf (0.1 Mcf) per month	$1.98/Mcf
Next	99 Mcf/month	4.82/Mcf
Next	1900 Mcf/month	4.35/Mcf
All over	2000 Mcf/month	4.20/Mcf

4.2. Given the following schedule for monthly electricity cost, calculate the monthly bill for a monthly energy use of 180,000 kwh, a demand (peak 15-min use rate) of 750 kw, and a power factor of 75%:

Cost per kwh: $.15 for first 500 plus 70 kwh/kw of demand in excess of 5 kw up to 450 kw
.12 for next 55 kwh/kw of demand in excess of 450 kw
.10 for next 10,000 kwh
.08 for all additional kwh

Power factor correction: The demand is to be calculated from the greatest 15-min peak demand rate during the billing period, plus .9% for every percent by which the power factor is less than 90%, with the final demand corrected to the nearest kw.

4.3. If the company in Problem 4.2 can increase its load factor to 65% by decreasing its peak with no decrease in total energy consumption, how much will the monthly bill be decreased?

4.4. How much is the monthly bill decreased by removing 1 kw from the electrical load (all the time) at the present energy consumption rates described in Problem 4.2?

4.5. In Problem 4.2, how much can be saved per year by increasing the power factor to 90%?

4.6. Calculate the electric bill for the monthly energy consumption of Problem 4.1 given the following electric schedule:

Demand charge	
First 40 kw of billing demand, per kw	$5.55
Next 460 kw of billing demand, per kw	4.53
Any additional kw of billing demand, per kw	3.32
Energy charge	
First 2500 kwh, per kwh	$ .056
Next 12,500 kwh, per kwh	.050
Any additional kwh, per kwh	.047
Power factor adjustment: Same as in Problem 4.2	

4.7. A church is typical of a firm with low utilization rates for much of the facility. A particular church has a gymnasium with 16 500-w incandescent ceiling lights. An equivalent amount of light could be produced by 16 250-w PAR (parabolic aluminized reflector) ceiling lamps. The difference in price is $5.00 per lamp, with no difference in labor. The gymnasium is used 9 months each year. How many hours per week must the gymnasium be used in order to justify the cost difference of a 1-year payback? Assume that the schedule used is that of Problem 4.6, that gymnasium lights do contribute to the peak demand (which averages 400 kw), and that the church consumes enough electricity that much of the bill comes from the lowest cost block in the table.

4.8. Find the equivalent constant-dollar after-tax present worth of the following 6-year project: purchase and installation cost, $100,000; maintenance per year, $10,000, increasing at 5%/year; energy saving per year, $45,000, increasing at 18%/year; salvage value, $20,000, increasing at 6%/year; the Consumer Price Index (CPI) projected to increase at 6%/year. Assume that the minimum attractive constant-dollar rate of return is 12%/year. Assume that the corporate tax rate is 48%, that an initial 10% credit is allowed for all investments with estimated service lives of 4 years or more, and that the equipment has a 5-year life for tax purposes.

Chapter 5

. . . .

Lighting

5.0 INTRODUCTION

Lighting is significant in energy management for two primary reasons: its psychological impact and its many opportunities for energy cost savings. Lighting provides one of the elements needed for most indoor work, and any uniform lighting policy thus affects most people who work indoors. As a starting place for an energy management program, lighting thus attracts immediate employee attention and can attract employee participation, since everyone has ideas about lighting. Lighting is also seen as a barometer of the attitude of top managers toward energy management: If the office of the president of a company is an example of wasteful lighting, then employees will assume that energy management is not taken seriously. Lighting also provides many opportunities for cost savings at low cost and little inconvenience. In this chapter we give a systematic way for finding those opportunities.

5.1 THE LIGHTING SYSTEM: FUNCTIONS AND COMPONENTS

Lighting is used primarily for workplace illumination, for safety, and for decoration. In each of these uses, the same three questions can be asked: (1) How much light is needed? (2) When is lighting necessary? (3) How can lighting be provided most efficiently? The answers can then be compared with the description of the existing system and the results used to decrease lighting cost and improve lighting efficiency.

5.1.1 Determining the Amount of Lighting Needed

Consider first question 1, how much light is needed. The Illuminating Engineering Society (IES) of North America has published standards giving ranges of recommended lighting levels some of which are shown in Tables 5–1 and 5–2 [1]. In using these tables, the Illuminating Engineering Society recommends that the lower values be used for occupants whose age is under 40 and/or where the room reflectance is greater than 70% and that the higher values be used for occupants more than 55 years old and/or where the room reflectance is less than 30%. For occupants between 40 and 55 years of age and where the reflectance is 30–70% or where a young occupant is combined with low reflectance or an older person is in a high-reflectance environment, the intermediate values should be used. In addition, the need for speed and accuracy influences the amount of light needed, with more critical speed and accuracy demanding more light.

In addition to the IES recommendations, the state of New Jersey has published a recommended standard of 1 w/ft² overall for a facility with a table showing additional standards for specific tasks. Part of this table is shown as Table 5–3.

Tables 5–1 through 5–3 provide guidelines for answering the first question, how much lighting is necessary. This answer must also include some definition of the needs for quality in light, defined in part by color rendition requirements. The color rendition of various light types is summarized in Table 5–4.

5.1.2 Determining When Lighting Is Needed

The answer to the second question—when is lighting needed—is that it is needed when workplace illumination, corridor lighting, or safety lighting is demanded by the presence of people or by security needs and when this light is not provided from natural sources. When lighting is not needed, lights should generally be turned off, and equipment for assuring this, discussed in Section 5.2.2.2, can yield quick payback of invested capital.

5.1.3 Cost and Performance Factors: Light Sources and Luminaires

The third question concerns providing adequate light at the least cost. Light sources differ substantially in their efficacy (ability to convert electricity into light) as shown in Table 5–5, and continual progress is being made within each category. In addition to light sources, however, a lighting system consists of luminaires, ballasts, and a schedule for maintaining and replacing luminaires, ballasts, and lights. Each one of these factors will influence the annual operating cost of the light system.

Ballasts are used in fluorescent and high-intensity discharge light sources to limit the current through the lamps and to provide starting voltages for these lights. Ballasts usually consume an additional 15–20% of the rated lamp wattage, a figure that is improving with competitive improvements in technology.

TABLE 5.1 ILLUMINANCE CATEGORIES AND ILLUMINANCE VALUES FOR GENERIC TYPES OF ACTIVITIES IN INTERIORS

| Type of activity | Illuminance category | Ranges of illuminances | | Reference work plane |
		Lux	Footcandles	
Public spaces with dark surroundings	A	20–30–50	2–3–5	General lighting throughout spaces
Simple orientation for short temporary visits	B	50–75–100	5–7.5–10	
Working spaces where visual tasks are only occasionally performed	C	100–150–200	10–15–20	
Performance of visual tasks of high contrast or large size	D	200–300–500	20–30–50	Illuminance on task
Performance of visual tasks of medium contrast or small size	E	500–750–1000	50–75–100	
Performance of visual tasks of low contrast or very small size	F	1000–1500–2000	100–150–200	
Performance of visual tasks of low contrast and very small size over a prolonged period	G	2000–3000–5000	200–300–500	
Performance of very prolonged and exacting visual tasks	H	5000–7500–10000	500–750–1000	Illuminance on task, obtained by a combination of general and local (supplementary lighting)
Performance of very special visual tasks of extremely low contrast and small size	I	10000–15000–20000	1000–1500–2000	

Source: Courtesy of the Illuminating Engineering Society of North America [1].

TABLE 5.2 LIGHTING RECOMMENDATIONS FOR SPECIFIC TASKS[a]

Area-activity	Illuminance category
Bakeries	D
Classrooms	D to E
Conference rooms	D
Drafting rooms	E to F
Hotel lobbies	C to D
Home kitchens	D to E
Inspection, simple	D
Inspection, difficult	F
Inspection, exacting	H
Machine shops	D to H
Material handling	C to D
Storage, inactive	B
Storage, rough, bulky items	C
Storage, small items	D
	Footcandles
Building entrances	1–5
Bulletin boards, bright surroundings, dark surfaces	100
Bulletin boards, dark surroundings, bright surfaces	20
Boiler areas	2–5
Parking areas	1–2

[a]For more detail, see Reference 4.

Source: Courtesy of the Illuminating Engineering Society of North America [1].

5.1.4 Cost and Performance Factors: Maintenance

In addition to a proper choice of light sources, ballasts, and luminaires, the efficiency of a lighting system depends on maintenance policies. Lighting system efficiency is degraded by lumen depreciation of the light sources, dirt accumulation on the luminaires, and dirt accumulation on the surrounding reflective surfaces.

5.1.4.1 Group relamping.

Lumen depreciation can be described by performance curves and can be countered by replacement. Performance curves, such as those in Figures 5-1, 5-2, and 5-3, show how light output is reduced as a function of ordinary usage. It should be noted that the life is measured in hours of *use* rather than *installed hours*. In addition to degraded performance of individual lights, light system performance is decreased by the burning out (*mortality*) of individual lamps. Typical mortality curves, showing the percent of lamps in service as a function of time, are given in Figures 5-4, 5-5, and 5-6.

To counter lumen depreciation, the maintenance procedure used is relamping. All can be replaced at one time (group relamping), lamps can be replaced as they burn out (spot relamping), or enough lamps can be replaced to provide minimal lighting

TABLE 5.3 ALLOWABLE LIGHTING POWER FOR SPECIFIC TASKS

Task	Allowable power (w/ft^2)
Corridors	.8
Conference rooms	
Critical seeing task	2.7
Conferring	1.0
Entrance foyer	1.2
Commercial kitchens	1.2
Garages	
Service, repair area	2.3
Service, traffic area	.8
Parking, traffic lanes	.4
Offices	
Drafting rooms	4.1–5.4
Detailed paperwork	4.1
Private offices	1.9
Warehouses	
Rough, bulky material	.3
Medium	.4
Fine	2.0
Assembly	
Large, easy seeing	1.0
Large, difficult seeing	1.2
Medium assembly	2.3
Foundries	2.0
Inspection	
Ordinary	1.2
Difficult	2.3
Material handling	
Packing, labeling	1.2
Picking stock	1.0
Loading	.8
Inside truck bodies and freight cars	.7

Source: Abstracted from *Energy Management Workbook for New Jersey's Industries.* Courtesy of the State of New Jersey, Department of Energy.

without replacing all of them. The choice of methods depends on costs. General Electric [4] has provided the following per unit cost formulas:
For spot replacement,

$$C = L + S \qquad\qquad (5\text{--}1)$$

For group relamping (using selected lamps as interim replacements),

$$C = \frac{L + G + (B \times S)}{I} \qquad\qquad (5\text{--}2)$$

For group relamping (with no interim replacements),

$$C = \frac{L + G}{I} \qquad\qquad (5\text{--}3)$$

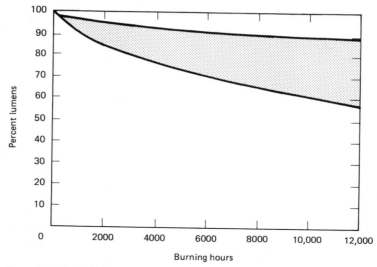

Figure 5–1 Typical lumen maintenance curve for flourescent lamps. (From General Electric Technical Pamphlet TP-105. Courtesy of General Electric Co.)

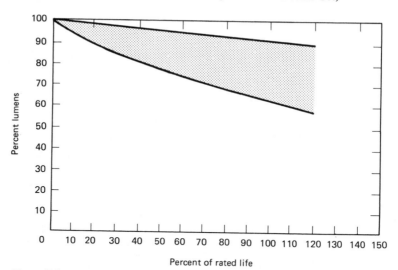

Figure 5–2 Typical lumen maintenance curve for filament lamps. (From General Electric Technical Pamphlet TP-105. Courtesy of General Electric Co.)

where C = total replacement cost per lamp
   L = net price per lamp
   S = spot replacement labor cost per lamp
   G = group replacement labor cost per lamp
   B = percent of burnouts at end of group relamping interval (from mortality curve)
   I = group relamping interval in percent of average lamp life

TABLE 5.4 COLOR RENDITION OF VARIOUS LAMP TYPES

	Filament[a]	Incandescent high-intensity discharge lamps				
		Clear Mercury	White Mercury	Deluxe White[b] Mercury	Multi-Vapor[b]	Lucalox[a]
Efficacy (lm/w)	Low	Medium	Medium	Medium	High	High
Lamp appearance effect on neutral surfaces	Yellowish white	Greenish blue-white	Greenish white	Purplish white	Greenish white	Yellowish
Effect on "atmosphere"	Warm	Very cool, greenish	Moderately cool, greenish	Warm, purplish	Moderately cool, greenish	Warm, yellowish
Colors strengthened	Red Orange Yellow	Yellow Green Blue	Yellow Green Blue	Red Yellow Blue Green	Yellow Green Blue	Yellow Orange Green
Colors grayed	Blue	Red, orange	Red, orange	Red	Red	Red, blue
Effect on complexions	Ruddiest	Greenish	Very pale	Ruddy	Grayed	Yellowish
Remarks	Good color rendering	Very poor color rendering	Moderate color rendering	Color acceptance similar to CW fluorescent	Color acceptance similar to CW fluorescent	Color acceptance approaches that of WW fluorescent

TABLE 5.4 (CONT.)

	Fluorescent lamps						
	Cool[b] White	Deluxe[b] Cool White	Warm[a] White	Deluxe[a] Warm White	Daylight	White	Soft White/Natural
Efficacy (lm/w)	High	Medium	High	Medium	Medium–high	High	Medium
Lamp appearance effect on neutral surfaces	White	White	Yellowish white	Yellowish white	Bluish white	Pale yellowish white	Purplish white
Effect on "atmosphere"	Neutral to moderately cool	Neutral to moderately cool	Warm	Warm	Very cool	Moderately warm	Warm pinkish
Colors strengthened	Orange Yellow Blue	All nearly equal	Orange Yellow	Red Orange Yellow Green	Green Blue	Orange Yellow	Red Orange
Colors grayed	Red	None appreciably	Red, green, blue	Blue	Red, orange	Red, green, blue	Green, blue
Effect on complexions	Pale pink	Most natural	Sallow	Ruddy	Grayed	Pale	Rudy pink
Remarks	Blends with natural daylight; good color acceptance	Best overall color rendition; simulates natural daylight	Blends with incandescent light; poor color acceptance	Good color rendition; simulates incandescent light	Usually replaceable with CW	Usually replaceable with CW or WW	Tinted source; usually replaceable with CWX or WWX

[a]Greater preference at lower levels.

[b]Greater preference at higher levels.

Source: General Electric Technical Pamphlet TP-119 [3]. Reprinted with permission of General Electric Co.

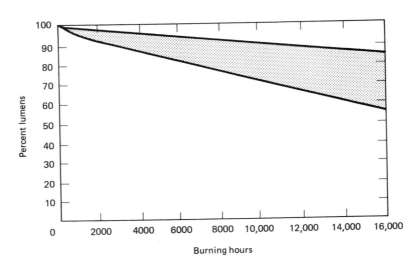

Figure 5-3   Typical lumen maintenance curve for HID lamps. (From General Electric Technical Pamphlet TP-105. Courtesy of General Electric Co.)

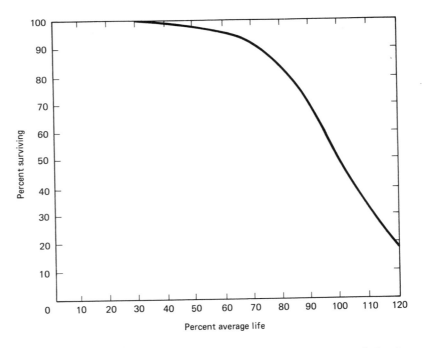

Figure 5-4   Typical mortality curve for flourescent lamps. (From General Electric Technical Pamphlet TP-105. Courtesy of General Electric Co.)

TABLE 5.5 EFFICACY AND USE OF VARIOUS LIGHT SOURCES

Light source	Wattage	Efficacy (lm/w)	Best use
Incandescent	15–1500	15–25	Domestic, seldom used spaces
Incandescent, PAR[a]		15–25	Displays, small area lighting
Fluorescent	15–219	55–100	Offices, low ceilings, task lighting, commercial uses
Mercury vapor	40–1000	50–60	Roadways, security, some indoor commercial
Metal halide	175–1000	80–100	Security, high ceilings, warehouses
High-pressure sodium	70–1000	75–140	Security, high ceilings, warehouses

[a]PAR: parabolic aluminized reflector.

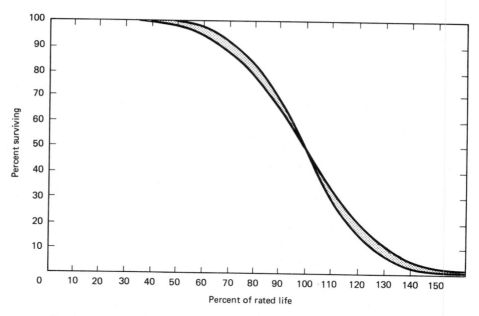

Figure 5-5 Typical mortality curve for filament lamps. (From General Electric Technical Pamphlet TP-105. Courtesy of General Electric Co.)

Example

An office building contains a number of smaller (400 ft²) rooms, each of which has four two-lamp fluorescent fixtures, and you must decide whether group relamping is preferable to spot relamping and, if so, the interval for relamping. Every time a janitor changes lamps, he must bring a ladder into the room and clear away furniture. If one lamp is replaced, this takes 15 min. If all the lamps are replaced, it takes 25 min. In addi-

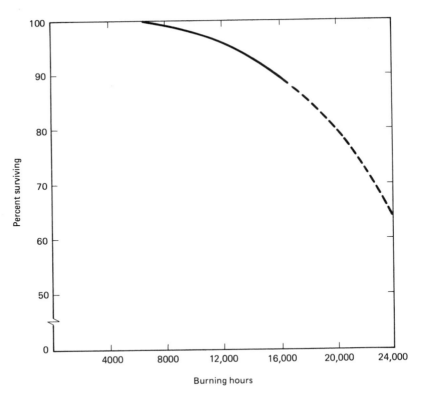

Figure 5-6 Typical mortality curve for HID (mercury vapor) lamps. (From General Electric Technical Pamphlet TP-105. Courtesy of General Electric Co.)

tion, it takes a total of 5 min/lamp to remove the existing lamp, clean the luminaire, and insert a new lamp. Lamps cost $.80 each, and labor costs $10.00/h. The office lights are on 8 h each day, 5 days each week, or approximately 2000 h each year. The average lamp life is 4000 h.

Solution.   In the preceding formulas, L = $.80/lamp, $S = \frac{20}{60} \times \$10.00$ or $3.33, and $G = \frac{30}{60} \times 1/8 \times 10.00 = \$.63$. If lamps are replaced every 6 months, the interval is approximately 1000 working hours long, I = .25, and Equations 5-1 and 5-3 yield $4.13 and $5.72/lamp for spot and group relamping, respectively. If the interval is increased to every year, I = .50, and these costs are $4.13 and $2.86, respectively. If group relamping is performed every year and is supplemented by interim replacement of failed lamps, the cost per lamp, from Equation 5-2, is ($.80 + $.63 + .50 × $3.33)/1 or $3.10/lamp. (It is assumed that each of these alternatives provides sufficient lighting, even at the end of the period.)

5.1.4.2 Luminaire maintenance.   In addition to replacing lamps, cleaning the surfaces of luminaires can be important. The amount of attention the luminaires will need depends on the kind of luminaire and the amount of dirt in the surrounding atmosphere. These factors have been quantified by the Illuminating Engineering Soci-

ety as follows: First find the maintenance category of the luminaire from Table 5-6. Then find the prevalent dirt condition from Table 5-7. With these, enter Figure 5-7 to find the percent of dirt depreciation on luminaires associated with a given maintenance interval.

5.1.4.3 Maintenance of reflective surfaces.

Dirt accumulating on reflective surfaces of a room can decrease the amount of reflected light substantially. A measure of the amount of light degradation is given by Figure 5-8, used with the courtesy of the Illuminating Engineering Society.

In addition to cleaning reflective surfaces, repainting dark-colored walls and ceilings with light colors adds substantially to the effective light within rooms and to the overall efficiency of the lighting system. Although it is difficult to quantify the improvement in changing from a darker room to a lighter room, the *IES Handbook* states the following:

> Light walls and ceilings, whether neutral or chromatic, are much more efficient than dark walls in conserving energy and distributing light uniformly. . . . Mathematical analyses . . . show that an increase of wall reflectance by a factor of 9 [e.g., from very dark to very light] could result in an increase in illuminance by a factor of about 3 [1, p. 5-21].

5.2 AUDITING THE LIGHTING SYSTEM

Having gained an understanding of the functions and components of a lighting system in Section 5.1, the person interested in energy management is now ready to determine where and how his or her lighting system can be made more efficient. The first step is to collect data to describe both the present lighting system and the needs of the present facility. This is followed by a careful survey of possible improvements to the system and the implementation of the most promising. The final step is the periodic monitoring of the lighting system to assure its continued efficiency.

5.2.1 Examining the Present Lighting System

To make recommendations to improve lighting system efficiency, it is necessary to know the present system. This system is defined when all the lights have been identified by type, wattage, and location; when the amounts of lumen depreciation, luminaire dirt, and reflective surface dirt have all been assessed; and when the lighting needs of each area have been listed. The facility is first divided into homogeneous areas. Then a lighting system condition form containing the information shown in Figure 5-9 is completed, using an industrial light meter to obtain the footcandle readings in each area and making judgments about the surface dirt conditions on the luminaires and the reflective surfaces. This form also requires calculation of the percentage of lamps that are no longer in service. Figure 5-10 is then used to define the actual light needs. To use Figure 5-10, take the lighting recommendations for

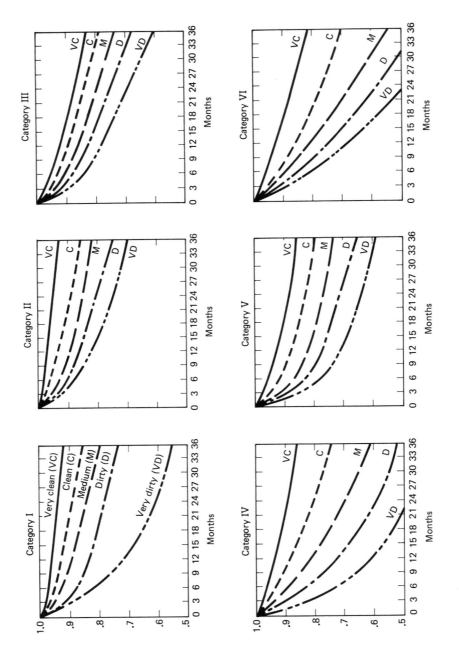

Figure 5-7 Luminaire dirt depreciation factors. (Courtesy of the Illuminating Engineering Society of North America.)

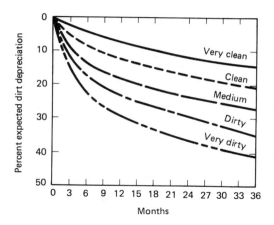

Figure 5-8   Dirt depreciation of reflective surfaces. (Courtesy of the Illuminating Engineering Society of North America.)

Location _____					Light type (HPS, FL, etc.) _____	

Lamps _____

Lighting level (fc)	No.	Watts		% burning	Condition of system*	
		Each	Total		Luminaires	Reflective surfaces
_____	__	__	__	____	___	___
_____	__	__	__	____	___	___
_____	__	__	__	____	___	___
_____	__	__	__	____	___	___
_____	__	__	__	____	___	___

*From Figure 5-7

Figure 5-9   Light system condition form.

Location	Hours when light is needed	Importance of color rendition	Task lighting possible ?	Light levels required (fc)
____	_____	_____	_____	_____
____	_____	_____	_____	_____
____	_____	_____	_____	_____
____	_____	_____	_____	_____

Figure 5-10   Lighting needs form.

each area from either Figure 5–1 or the *IES Handbook* and list them on the form. Determine when the light is actually needed (and determine for yourself that the need is real) and write in these hours. Also note whether it is important to have good color rendition and whether it is feasible to use local lighting focused directly on the workplace (task lighting) rather than area lighting.

TABLE 5.6   MAINTENANCE CHARACTERISTICS OF LUMINAIRES

Maintenance category	Top enclosure	Bottom enclosure
I	1. None	1. None
II	1. None	1. None
	2. Transparent with 15% or more uplight through apertures	2. Louvers or baffles
	3. Translucent with 15% or more uplight through apertures	
	4. Opaque with 15% or more uplight through apertures	
III	1. Transparent with less than 15% upward light through apertures	1. None
		2. Louvers or baffles
	2. Translucent with less than 15% upward light through apertures	
	3. Opaque with less than 15% uplight through apertures	
IV	1. Transparent unapertured	1. None
	2. Translucent unapertured	2. Louvers
	3. Opaque unapertured	
V	1. Transparent unapertured	1. Transparent unapertured
	2. Translucent unapertured	2. Translucent unapertured
	3. Opaque unapertured	
VI	1. None	1. Transparent unapertured
	2. Transparent unapertured	2. Translucent unapertured
	3. Translucent unapertured	3. Opaque unapertured
	4. Opaque unapertured	

Source: Courtesy of the Illuminating Engineering Society of North America.

Another step in reducing energy costs associated with lighting is the midnight audit. A midnight audit is conducted when most operating personnel have left the area, at night, in the company of security personnel. (Security personnel have the keys and are responsible for the safety of people and equipment at night. They are also a source of ideas and can implement many changes.) In this audit, look for unnecessary lighting, listen for motors that may be on unnecessarily, and listen for steam and air leaks whose noise may be hidden during the day. Record all suggestions.

5.2.2 Finding and Implementing Improvements

The lighting surveys define the present system and the actual needs of the facility. The next step is to use the survey information to search for possible improvements. Three categories should be considered to start with: relamping, changing the control system, and improved maintenance. Consider each of these in turn.

5.2.2.1 Relamping.   If the lighting survey has shown that a room has more light than is needed, the room is a candidate for relamping—eliminating some existing lamps, replacing existing lamps with lamps that are more efficient, and replacing existing ballasts with more modern ballasts that provide as good or better service and use less power.

TABLE 5.7 DEGREES OF DIRT CONDITIONS

	Very clean	Clean	Medium	Dirty	Very dirty
Generated dirt	None	Very little	Noticeable but not heavy	Accumulates rapidly	Constant accumulation
Ambient dirt	None (or none enters area)	Some (almost none enters)	Some enters area	Large amount enters area	Almost none excluded
Removal or filtration	Excellent	Better than average	Poorer than average	Only fans or blowers if any	None
Adhesion	None	Slight	Enough to be visible after some months	High—probably due to oil, humidity, or static	High
Examples	High-grade offices, not near production; laboratories; clean rooms	Offices in older buildings or near production; light assembly; inspection	Mill offices; paper processing; light machining	Heat treating; high-speed printing; rubber processing	Similar to dirty but luminaires within immediate area of contamination

Source: Courtesy of the Illuminating Engineering Society of North America.

By far the easiest measure is to eliminate some existing lamps. Remove lights to the point where existing lighting is minimally sufficient for the use of the room, making sure to remove both lamps that are on the same ballast in any given fixture. The lamps that have been removed can then be used for replacement, thus saving maintenance dollars in addition to those saved in energy. When the lamp removed is fluorescent, the ballast should also be removed. In addition to saving the 15–20% of energy that the ballast adds to the energy consumption of the lamp, this measure makes available more ballasts for replacement, again with substantial savings.

A second measure is to replace existing lamps and fixtures with lamps and fixtures that are more efficient. The business of selling lamps, luminaires, and ballasts is intensely competitive, and new, more efficient products are being developed all the time. Some useful substitutions are given in Table 5-8 [2].

A final measure that should not be overlooked is to replace existing ballasts with ballasts that are more energy efficient. A ballast usually consumes an additional 15–20% of the energy of the lamp it supports. As the competitive process leads to more efficient technology, this figure is approaching 10–12% and will certainly go below that. In addition, the life of ballasts is becoming longer as cooler units are developed, thus lowering the cost per unit of this component of the lighting system. Reliable vendors are claiming an 8-month payback (before taxes) on ballast replacements, so it is an option worth considering.

5.2.2.2 Control system measures.   There is no automatic reason for supplying more light than a given task or area requires or to supply the light when an area is not in use. An ideal control system would ensure that the amount of light is kept to a minimum at the same time that lights are on only when lighting is needed. Most existing control systems do not approach this ideal, but there are some reasonable measures that can be taken to substantially improve their operation: (1) Change the span of the control system; (2) install controls to turn lights on only when lighting is needed and to turn them off when the need has passed, and (3) install controls or procedures to take advantage of natural outside light whenever possible.

The *span* of the lighting control system is the number of lights that are controlled from one switch. If this number is very large, it is necessary to illuminate an entire area in order to provide light for anyone in the area, and the lighting level of the entire area is determined by the single switch. It is clearly undesirable to control all the lights in a building from one switch; it may be equally undesirable to have all lights on one floor or in one room so controlled. At least two alternatives are possible: Install switches so that the lights can be turned on at the point of use, or connect every second light to one switch so that the usual lighting condition is for only half of the lights to be on. Either way saves money.

One way to illuminate areas only when people are present or security lighting is needed is with the use of timers preset to the working schedule of the facility or to the time when the lighting is essential for security. If lighting is used for decorative purposes only, it would seem obvious to turn it off during the hours when no one is around to observe the decoration. If lighting is used for security, a photoelectric cell can be used to turn the lights on when night comes, and a timer can be used to turn

TABLE 5.8 LAMP SUBSTITUTIONS[a]

Present lamp	Substitute	Light level	Energy saved (w / %)
60-w In (1000 h)	30-w RI 50-w RI	100% 200%	30/50 10/16
60-w In (2500 h)	50-w RI 55-w PAR/FL	200 + % 200 + %	10/16 5/8
75-w In	55-w PAR/FL	150%	20/27
100-w In (750 h)	75-w IR 75-w PAR/FL 75-w ER	125% 200 + % 200 + %	25/25 25/25 25/25
100-w In (2500 h)	75-w PAR/FL	200 + %	25/25
150-w PAR/FL	100-w PAR/FL	70%	50/33
150-w R/FL	100-w PAR/FL	150%	50/33
200-w In (750 h)	150-w PAR/FL (2000 h)	200 + %	50/33
30-w F	30-w EEL	87%	5/16
40-w F	40-w EEL	89%	6/15
96-w F	96-w EEL	91%	15/20
96-w F/HO	96-w EEL/HO	91%	15/14
96-w F/SHO	96-w EEL/SHO	90%	30/14
175-w MD	100-w HPS	104%	75/42
400-w MD	200-w HPS	96%	200/50
300-w In	150-w HPS	250%	120/50
750-w In	150-w HPS	104%	570/80
1000-w In	200-w HPS	93%	770/80

[a] Abbreviations:

EEL,	energy-efficient fluorescent light (such as Matt-Mizer, Super-Saver, etc.)
ER,	elliptical reflector (shape and inside coating of lamp)
F,	fluorescent
FL,	floodlight
HO,	high output: 1000-ma filament
HPS,	high-pressure sodium
h,	hour (mean life expectancy)
In,	incandescent
MD,	mercury deluxe: mercury vapor corrected to improve color
PAR,	parabolic aluminized reflector (see ER)
RI,	reflective coated incandescent
SHO,	superhigh output: 1500-ma filament
w,	watt

them off late at night at a time agreeable to the security personnel. If lighting is needed only when people are present, such as in a classroom, it may be possible to find controls that use detectors or counters to determine when people are present and turn the lights off at other times. Many of these opportunities present themselves during the midnight audit described, and most have short payback periods.

Natural light is a resource that can be used to save lighting dollars, usually inexpensively. If an area receives some direct light during the day, it is often possible to turn off lights there during the daytime or to install photoelectric cells to perform this task automatically. A new development has been the invention of dimming controls that control the light supplied by electricity so that the level of light inside the room remains constant. (There may be undesirable effects on light system component performance—check with lamp vendors before installing such a system.)

5.2.2.3 Lighting system maintenance. In Section 5.1.4.3 we described the effects of lumen depreciation and dirt accumulation on the lighting system, and Figures 5–4 through 5–8 gave the amount of light depreciation to be expected under various dirt conditions. It is clear from those figures and the associated written material that maintenance has a definite effect on lighting system performance. Instituting a lighting system maintenance program can thus be another way to improve delivered light performance; the lack of such a program can degrade the performance of a lighting system as surely as not replacing burned out lights.

5.3 SUMMARY

In this chapter we have discussed the lighting system and have given ways of improving its performance. The importance of the lighting system justifies giving major attention to it.

REFERENCES

1. *IES Lighting Handbook,* Illuminating Engineering Society of North America, New York, 1981.
2. *Energy Management Workbook for New Jersey's Industries,* State of New Jersey Department of Energy, Newark, N.J., 1981.
3. *Light and Color,* General Electric Technical Pamphlet TP–119, General Electric Co., Cleveland, Ohio, Feb. 1978.
4. *Lighting Maintenance,* General Electric Technical Pamphlet TP–105, General Electric Co., Cleveland, Ohio, Jan. 1969.

BIBLIOGRAPHY

HARROLD, RITA M., Lighting, in *Energy Management Handbook,* Senior ed. W. C. Turner, Wiley-Interscience, New York, 1982, Chap. 13.

QUESTIONS

5.1. How does lighting affect worker performance?

5.2. How much light is needed for roadway safety at night, and how should it be provided?

5.3. What are some factors, in addition to those discussed, that have an important impact upon the decision to spot- or group-relamp?

PROBLEMS

5.1. In auditing a plant during the daytime, you note 12 security lights on. Each light uses 280 w, including the ballast. One way to save some of the electricity cost being wasted is to install photocells, at an average cost of $15.00/light, including labor. The demand peak is during the daytime for the rest of the plant. Electricity costs $.08/kwh with a demand charge of $7.50/kw/month. The power factor is not a problem. What is the payback on this improvement?

5.2. A large open area in a factory is illuminated continuously with 225-w (including the ballast) fluorescent tubes, two per fixture, with 325 fixtures in the area. This area is approximately 150 × 300 ft, with lights 24 ft from the floor. Design a new lighting system using 250-w high pressure sodium lamps, and calculate the payback on the new system. Base your estimate of the number of lamps required on the efficacy alone, the sodium lamps having 2.5 the efficacy of the fluorescents. Assume that electricity costs $.045/kwh with a monthly demand charge of $6.30/kw, that lamps cost $10.00 each, and that removal of old fixtures and replacement with new fixtures costs $130/fixture. Assume that the mean life for fluorescents is 1.37 years and 2.74 years for the high-pressure sodium lamps.

5.3. In the situation in Problem 5.2, a detailed analysis was performed by a lighting contractor using the zonal cavity method prescribed by the Illuminating Engineering Society. This analysis showed that only 220 of the new lamps were needed. If you answered Problem 5.2, evaluate the initial and 1-year energy savings obtained by using better advice. If you didn't answer Problem 5.2, calculate the initial cost and first-year savings when comparing the better recommendation against the present system.

5.4. Consider a situation in which HID lamps have been installed in maintenance category III luminaires in an environment classed as dirty, the lamps have a 15,000-h mean life, luminaire cleaning is performed once each year (using compressed air in a long aluminum pipe), and group relamping is performed every 2 years with luminaire cleaning. Also assume that the lighting system operates 7500 h/year. What is the average delivered percentage of the original light level?

5.5. You have been asked to recommend whether spot relamping or group relamping should be used for the fluorescent lamps in a particular facility. Because of the height of the lamps, setup and take-down time for the ladders are $\frac{1}{2}$ h each. The time to replace a lamp and to clean each fixture (two lamps) is 10 min. The largest number of fixtures that can be relamped for any one setup and take-down is 42. One person is needed, at $12.00/h (including overhead). The lamps cost $1.50 each. Assuming that the group relamping interval would be 75% of rated life, is group relamping preferred over spot relamping, and, if so, by how much per lamp?

Chapter 6

Heating, Ventilating, and Air Conditioning

6.0 INTRODUCTION

The heating, ventilating, and air conditioning system (from now on referred to as the HVAC system) is the system of motors, ducts, fans, controls, and heat exchange units which delivers heated or cooled air to various parts of a building and exhausts the air from conditioned spaces. Air absorbs heat from lights, people, industrial processes, and the sun, and air conditioning removes the excess heat in order to provide a comfortable working environment. This system also adds heat if the working environment is too cold for worker comfort. Other conditions that necessitate air conditioning include contaminants in the air and excess humidity; it may also be necessary to air-condition an area to meet unusual requirements (such as those in a laboratory or a clean room) or to protect products.

There are many kinds of HVAC systems, depending on the fluid that is used as a heat exchange medium (usually water or air), on the particular requirements for the system, and on the type of system in style when the building was originally built. All HVAC systems have certain components in common: a source of heat, means for transferring the heat from the point of generation to the point of use, and a control system. The source of heat is usually a boiler, a furnace, or the sun. If cooling is used, the source of cold temperature is usually a chiller, although cold air can be supplied either as the exhaust air from a cold area or as air brought in during the early morning to flush out the building. The heat or cold is transferred from a furnace, a boiler, or a chiller to air, and this air is distributed to the points of use. An alternative system may distribute hot water to the points of use where it warms up air to be blown into the room. The control system may be as simple as a thermostat that turns on a furnace when it senses room temperature below a preset level, or it may be very elaborate, controlling air volume, humidity, and temperature through monitoring inputs from many sensors and actuating valves, motors, and dampers. In this chapter we analyze heating and cooling loads and ventilation requirements, describe components of HVAC systems, and give methods of improving the energy consumption characteristics of an existing system.

6.1 HEATING, COOLING, AND VENTILATING LOADS

One of the easier ways to reduce costs in HVAC systems is to reduce the amount of energy that must be added to or extracted from an area to bring the area to within a desired temperature range. Two major strategies are available: (1) Reduce the heating or cooling load; (2) change the targeted temperature range. The amount of *cooling* needed in an area can be reduced by reducing the amount of heat brought into the area through the walls, by reducing the number of people present, or by modifying the energy consumption characteristics of industrial processes. The amount of *heating* needed can be reduced by increasing the amount of machinery located indoors, by capturing some of the heat from lights, by reducing infiltration of cold outside air, or by insulating roofs or ceilings so that less heat escapes. In any of these examples, the cooling or heating load is being changed.

The second strategy is to change the temperature range labeled by the management as desirable, i.e, changing the temperature limit above which air cooling occurs, the upper set point, and the temperature limit below which heating occurs, the lower set point. If heating does not start until the temperature is 55°F or lower, less heat will be used than if the threshold temperature for heating is 65°F. Similarly, an upper control limit of 85°F is more economical than a limit of 75°F. Changing these temperatures has the effect of changing the heating or cooling load imposed upon the HVAC system, although none of the heat sources are changed.

6.1.1 Heating and Cooling Load Calculations

The heating and cooling loads in a building can be categorized into (1) heat given off by people; (2) radiant energy from the sun that enters through windows, is absorbed by furniture, walls, and equipment within the building, and is later radiated as heat within the building; (3) heat conducted through the building envelope (walls, roofs, floors, and windows) to or from the environment around the building; (4) waste heat given off by processes and machinery within the building; (5) heat given off by lighting; and (6) heat or cooling lost to ventilation or infiltration air. A methodology for calculating these heat and cooling losses and their interactions is presented in Reference 1 in sufficient detail that it can be used for design purposes. The intent of this book, however, is not to enable a person to design a new building but rather to enable him or her to manage the energy costs of an existing building; thus in this section we emphasize those aspects of the heating and cooling load that can be changed by moderate or low expense or by scheduling.

6.1.1.1 Heating and cooling load: people. People give off heat, and the amount they give off depends on the type of work they are doing, whether they are men or women, and the temperature of their surroundings. Table 6–1 gives representative values for the heat given off under various conditions. If no cooling or heating takes place during nonworking hours, the figures in Table 6–1 can be used directly. If cooling or heating takes place when the work force is not present, the later reradiation of heat given off by people and absorbed by equipment and surroundings must be taken into effect as described in the ASHRAE Fundamentals Handbook [1].

People-generated heat can be managed in several ways. The first management technique is that of scheduling: Descreasing the number of people in an air-conditioned area during the time of peak energy consumption will decrease the amount of heat that must be removed and thus will decrease this component of the peak demand. If the work of people can be scheduled when the outside tempeature is lower than the inside temperature, it is often possible to remove people-generated heat by introducing colder outside air rather than by using refrigerated air conditioning. Another technique is to remove people. This is accomplished by replacing people by automated equipment such as automated storage and retrieval systems. Removing the people decreases the cooling load and may completely eliminate the necessity for cooling (or heating) an area.

6.1.1.2 Heating and cooling load: solar radiation. The cooling load due to solar radiation through windows can be calculated by

$$q = \Sigma A \times SC \times MSHG \times CLF \qquad (6\text{–}1)$$

where A = window area (square feet)
  SC = shading coefficient
  MSHG = maximum solar heat gain
  CLF = cooling load factor

and the summation is over all surfaces of the building envelope. In Equation 6–1, A comes from measurement. The shading coefficient SC depends on the kind of glass, the indoor shading (roller blinds or venetian blinds), and the average outside wind velocity. For single-pane glass, typical values range from .53 for $\frac{1}{2}$-in. heat-absorbing glass to 1.00 for $\frac{1}{8}$-in. clear glass; a white roller shade inside a clear $\frac{3}{8}$-in. glass pane gives a shading coefficient of .25. Additional values of the shading coefficient can be obtained from glass vendors and from Reference 1. Changing glass types and installing shades are energy management measures that can make a significant change in this factor. The maximum solar heat gain and the cooling load factors, however, are fixed by the location and design of the structure. The maximum solar heat gain (MSHG) depends on the month, the hour, the latitude, and the direction the surface is facing. For 40° north latitude (Denver and Salt Lake City), these factors are shown in Table 6–2. The cooling load factor (CLF) measures the fraction of energy absorbed at a given time that is radiated as heat at a later time. This factor, like the shading coefficient, depends on the interior shading. Representative values for this factor are given in Table 6–3 for a building with medium-weight construction (ap-

TABLE 6.1 RATES OF HEAT GAIN FROM PEOPLE

Activity	Total heat gain for male adults (Btu/h)
Seated at rest	400
Seated, writing	480
Seated, typing	640
Standing, light work or slow walking	800
Light bench work	880
Normal walking, light machine work	1040
Heavy work, heavy machine work, lifting	1600

Note: Heat gain from adult females is assumed to be 85% of that for adult males.

Source: From *1977 Fundamentals Handbook and Product Directory,* © 1977. Reprinted with permission from the American Society of Heating, Refrigerating, and Air Conditioning Engineers, Atlanta, Ga.

proximately 70 lb of building material per f² of floor area). Additional values for these factors can be found in Reference 1.

Equation 6–1 is useful in managing the solar component of the cooling load. For buildings in a hot climate, it is often cost effective to cover all outside windows and thus reduce the area (A) through which solar radiation enters the building. If, however, heating is more of an expense than cooling, it may be possible to increase the area and use passive solar heating. (For more detail on the possibilities of this approach, see Reference 2.) The shading coefficient (SC) can be changed by the installation of shades, heat-absorbing glass, or reflective coatings. The maximum solar heat

TABLE 6.2 MAXIMUM SOLAR HEAT GAIN FACTORS
(Btu/h•ft² for 40° NORTH LATITUDE)

	Surface orientation					
	N	NE/NW	E/W	SE/SW	S	Horizontal
Jan.	20	20	154	241	254	133
Feb.	24	50	186	246	241	180
March	29	93	218	236	206	223
April	34	140	224	203	154	252
May	37	165	220	175	113	265
June	48	172	216	161	95	267
July	38	163	216	170	109	262
Aug.	35	135	216	196	149	247
Sept.	30	87	203	226	200	215
Oct.	25	49	180	238	234	177
Nov.	20	20	151	237	250	132
Dec.	18	18	135	232	253	113

Source: From Reference 1. Reprinted with permission from the American Society of Heating, Refrigerating, and Air Conditioning Engineers, Atlanta, Ga.

TABLE 6.3   COOLING LOAD FACTORS FOR GLASS

Solar time									
	N	NE	E	SE	S	SW	W	NW	Hor.

Solar time	N	NE	E	SE	S	SW	W	NW	Hor.
Without Interior Shading									
2 a.m.	.20	.06	.06	.08	.11	.13	.13	.12	.14
4	.16	.05	.05	.06	.08	.10	.10	.09	.11
6	.34	.21	.18	.14	.08	.09	.09	.09	.11
8	.46	.44	.44	.38	.14	.12	.10	.11	.24
10	.59	.40	.51	.54	.31	.15	.12	.14	.43
12	.70	.33	.39	.51	.52	.23	.14	.17	.59
2 p.m	.75	.30	.32	.40	.58	.44	.29	.21	.67
4	.74	.26	.26	.33	.47	.58	.50	.42	.62
6	.79	.21	.21	.25	.36	.53	.55	.53	.47
8	.50	.15	.15	.18	.25	.33	.33	.32	.32
10	.36	.11	.11	.14	.18	.24	.23	.22	.24
12	.27	.08	.08	.10	.14	.18	.17	.16	.18
With Interior Shading									
2 a.m.	.07	.02	.02	.03	.04	.05	.05	.04	.05
4	.06	.02	.02	.02	.03	.04	.04	.03	.04
6	.73	.56	.47	.30	.09	.07	.06	.07	.12
8	.65	.74	.80	.74	.22	.14	.11	.14	.44
10	.80	.37	.62	.79	.58	.19	.15	.19	.72
12	.89	.27	.27	.49	.83	.38	.17	.21	.85
2 p.m	.86	.24	.22	.28	.68	.75	.53	.30	.81
4	.75	.20	.17	.22	.35	.81	.82	.73	.58
6	.91	.12	.11	.13	.19	.45	.61	.69	.25
8	.18	.05	.05	.07	.09	.12	.12	.12	.12
10	.13	.04	.04	.05	.07	.08	.08	.08	.08
12	.09	.03	.03	.04	.05	.06	.06	.06	.06

Note: Solar time = local standard time (LST) + 4 min × (LST meridian − local longitude) + correction for the month (from − 13.9 min in February to + 15.4 min in October).

Source: From Reference 1. Reprinted with permission from the American Society of Heating, Refrigerating, and Air Conditioning Engineers, Atlanta, Ga.

gain (MSHG) can be used to help decide where to locate people and equipment within a building relative to the location of windows. More heat will come in if the windows are facing south than if the windows are facing north. If a close visual contact with the outside is desired but the sun heat is not, both conditions can be met with windows facing the north. The cooling load factor is reduced by shading, since less heat is absorbed by furniture and walls to be reradiated later.

6.1.1.3 Heating and cooling load: conduction.
In addition to inside heating caused by radiation of solar energy absorbed by inside materials, the sun heat combines with the outside temperatures to create heating due to conduction through the

walls and the roof. The heat conduction equations, though interesting, are beyond the scope of this text. Several observations can be made that are relevant to energy management, however. First, the amount of heat gain or loss through a wall depends on the construction of the wall as well as on the *U* value. The so-called *flywheel effect* describes this: In the same way that a flywheel at motion tends to remain in motion, a warm wall tends to remain warm and to radiate heat after the sun is down. The amount of heat and the time the wall continues to radiate depend on the construction of the wall in the same sense that the speed and running time of a released flywheel depend on the construction of the wheel. Second, adding insulation to walls or roofs can be very cost effective if the main component of the cooling load is from conduction rather than from inside sources (insulation is discussed in detail in Chapter 11). Third, if an existing building has massive walls, it may be possible to schedule work hours so as to take advantage of the flywheel effect and obtain free heating or cooling. These three points are discussed in Reference 3.

6.1.1.4 Heating and cooling load: equipment.

The fourth major source of heating is from equipment. The Btu per hour from ovens, industrial processes such as solder pots that use much heat, and many other types of equipment, can either be read directly from nameplates or can be approximated from gas or electricity usage by assuming that every kwh of electricity contributes 3412 Btu of heat at the point of use and that each Mcf (thousand cubic feet) of gas has an energy content of 1 million Btu. The amount of heat given off by a motor can also be estimated from the voltage used by the motor. If it is a two-phase motor, the energy converted to heat per hour is given by

$$\text{Btu/h} = \text{voltage} \times \text{current} \times \frac{3412 \text{ Btu/kwh}}{1000 \text{ w/kw}} \qquad (6\text{-}2)$$
$$\times \text{ use factor (fraction of time that motor is in use)}$$
$$\times (1 - \text{motor efficiency})$$

For a three-phase motor, the energy converted to heat per hour is given by

$$\text{Btu/h} = \text{voltage} \times \text{current per phase} \times 1.732 \times 3.412 \qquad (6\text{-}3)$$
$$\times \text{ use factor} \times (1 - \text{efficiency})$$

(The efficiency for a single-phase motor is usually 50–60%, and the efficiency for a three-phase motor is usually 60–95%.) The energy used by other kinds of equipment can be found in Reference 1, from equipment vendors, and from gas and electric utilities.

Equations 6-2 and 6-3 show that decreasing the amount of equipment using electricity or gas at a given time has a twofold effect. First, the actual energy used at that time is decreased. Second, the heat introduced into the space is also reduced. If this heat must be removed by cooling, the cooling load is thus reduced by turning off this equipment. If this heat is desirable, Equations 6-2 and 6-3 can be used to indicate how much waste heat can be made available by equipment use scheduling.

6.1.1.5 Heating and cooling load: lighting. Any energy that is used in lighting ultimately becomes part of the cooling load, so any reduction in lighting automatically reduces the amount of heat to be removed.* To calculate the reduction in heat removal, remember that each kwh of electricity is equivalent to 3412 Btu. Thus the decrease in heat burden is given for each hour by

$$\text{Btu/h} = 3412 \times K \tag{6-4}$$

where K is the number of kw by which the lighting energy has been decreased. Note that K must include the energy consumed by any ballasts that are connected to the lights. To obtain the total energy saving, Equation 6-4 is calculated for each hour and the total obtained.

Other ways of reducing the amount of lighting energy needed to illuminate a space to a prescribed level include replacing bulbs with more efficient bulbs, cleaning the luminaires, and painting adjacent surfaces. These and other measures were discussed in detail in Chapter 5.

6.1.1.6 Heating and cooling load: air. The sixth major category of heating or cooling load comes from energy used to heat, cool, or humidify air. This part of the heating or cooling load can be reduced by weatherstripping, caulking, and tightening windows; by installing loading dock shelters; by replacing broken windows; and by other measures designed to reduce or eliminate air leakage to or from the outside. Air infiltration can occur through the envelope at many places, and there are infiltration-reducing techniques unique to each place. Such techniques range from caulking to dock seals (discussed in Section 2.1.2.1). It can also prove worthwhile to prevent airflow from conditioned areas of a plant to unconditioned areas (or vice versa) by creating airflow barriers indoors such as the plastic curtain shown in Figure 6-1.

It is desirable to be able to estimate the cost or benefit of any measure. Infiltration or exfiltration generally involves air moving through openings. The annual amount of energy lost through a hole or crack can be estimated from

$$\text{Btu/year} = V \times 1440 \times .075 \times (.24 + .45W) \times (\text{HDD} + \text{CDD}) \tag{6-5}$$

where V = volume of air entering or leaving, in cubic feet per minute
 1440 = number of minutes per day
 .075 = pounds of dry air per cubic foot
 .24 = specific heat of air, in Btu per pound per degree F
 .45 = specific heat of water vapor, in Btu per pound per degree F
 HDD = heating degree days per year, in days × degrees F
 CDD = cooling degree days per year, in days × degrees F

(This formula is a modification of that given in Reference 1). The impact of water

---

*Some heating systems have been designed to use lighting as a heat source. If your building has such a system, be careful about reducing the amount of lighting.

Figure 6–1 One air leakage remedy: plastic curtain for lift trucks. (Courtesy of Flexion, Inc.)

vapor is ignored in Equation 6-5 because exhausting it is a benefit for cooling and a disbenefit for heating and, in either case, the effect is usually negligible.

6.1.2 Ventilation Requirements for Health

In addition to satisfying the need for a comfortable working environment, the HVAC system must also provide ventilation to remove noxious substances from the air. Many ventilation requirements are specified by state and local health and safety codes. An example of such requirements is given in Table 6–4. Meeting this kind of requirement imposes some demands on the creativity of any person trying to reduce energy costs. Some of the methods that have worked to meet these requirements without the necessity for heating excessive amounts of outside air have included the installation of special equipment such as self-ventilating hoods and isolating contaminant areas from other parts of a plant.

Another problem is created by the need to get rid of cigarette smoke. Where people are not smoking, the recommended standard is 5-7.5 cfm/person; where smoking is permitted, the recommended standard is 25-40 cfm. In this situation, the amount of ventilation air is reduced by having separate areas for smokers and non-smokers, a separation which is enforced by law in many states.

TABLE 6.4 BREATHING AIR REQUIREMENTS

Environment	Minimum (cfm)	Recommended (cfm)
Residences (single-family)		
General living areas	5	7–10
Kitchens, bathrooms	20	30–50
Commercial		
Shipping and receiving areas	15	15–20
Warehouses	7	10–15
Bars	30	40–50
Conference rooms	25	30–40
Enclosed parking garages (per ft^2)	1.5	2–3
General office space	15	15–25
Institutional		
Classrooms	10	10–15
Hospital foyers	20	25–30
Animal research rooms	40	45–50

Note: Many ventilation requirements are set by local codes.

Source: From *1977 Fundamentals Handbook and Product Directory,* © 1977. Reprinted with permission from the American Society of Heating, Refrigerating, and Air Conditioning Engineers, Atlanta, Ga.

6.1.3 Ventilation Standards for Comfort

One of the main reasons for having air conditioning is to keep people comfortable. The definition of comfortable conditions changes with clothing styles, the number and quantity of local drafts, and the price of energy for heating, cooling, and humidifying. Comfort is generally defined, however, by a temperature range that varies depending on the relative humidity of the air. Three measures take advantage of the comfort zone: lowering the minimum temperature at which heating is initiated, raising the maximum temperature at which cooling is initiated, and changing the humidity in the air to more nearly conform to that of the outside environment. If outside air can be used without either humidifying or dehumidifying, considerable energy can be saved. (The exact amount depends on the HVAC system, but note that evaporating 1 lb of water uses 1040 Btu. The method of cooling air so that condensation takes place and then reheating the dried air to a comfortable temperature is also costly.) Another method for saving energy is to reduce the amount of ventilation air to that required by the given environment; the cube law for fan horsepower can be used to estimate possible savings. Note that management would bear the burden of selling these measures to the employees or of introducing the changes so gradually that they are not noticed.

Another method of energy management that relates to comfort is to allow temperatures to slowly change. Several articles in the *ASHRAE Journal* have stated that people are relatively insensitive to slow changes in air temperature [4, 5]. This insensitivity can be used to advantage by turning the air conditioning down or off 30 min or an hour before quitting time and letting the temperature drift.

6.2 HOW AN HVAC SYSTEM WORKS

If no one involved in energy management in a firm understands how the firm's HVAC system works, then the chances are excellent that the system is not being operated correctly. Our purpose in this section is to enable the reader to understand and analyze the operation of his or her HVAC system. In Section 6.3 we then give some methods of improving the operation of such systems.

6.2.1 A Dual-Duct System

6.2.1.1 Operation.   One way to understand how HVAC systems work is to first learn how one system works and then to see other systems as variations of that system. One of the more widely used HVAC systems is the dual-duct system, illustrated in Figure 6-2. In this system, outside air enters through dampers, is mixed with return air in a mixing box or plenum, and is forced by a supply fan through a cooling unit or through a heating coil. The air that has been cooled passes through a cold supply duct to a mixing box and thence into a room. Air that has been heated goes through a hot air duct to the mixing box. The proportion of cold and hot air that are mixed together determines the temperature of the air supplied to the room; in addition, there is usually a control to regulate the volume of air entering the room.

As new air is entering a room, used air is being exhausted from the room. This air leaves through a return air grille and is drawn through the return air duct by a return air fan. Some of the return air is exhausted to the atmosphere, and some is mixed with incoming outside air for use as supply air for rooms.

The dual-duct system has the advantage that it can accommodate widely differing demands for heating and cooling in different zones of a building by changing the ratio of hot to cold air in each zone. Its control system is also easily understood and relatively maintenance-free. The dual-duct system has the disadvantages that, first, it requires much ductwork with attendant cost and space usage, and, second, when the hot and cold ducts are next to each other, unproductive heat transfer takes place.

6.2.1.2 System components.   The components of this system are dampers, grilles, filters, coils, fans, ductwork, and the control system. Each of these contributes to the operation of the HVAC system as follows.

Dampers.   A damper controls a flow of air. If the damper is open, the air can flow unimpeded; if it is closed, the flow is reduced to 5–10% of open-damper flow, with the percentage dependent on the construction and maintenance of the damper. Dampers are usually used to regulate the flow of air from outside into a system or to control the flow from one part of the system into another part, as in the case of a return air damper (A in Figure 6-2). In Figure 6-2, if the return air damper is closed and the outside air damper open, all the heat (or cooling) in the return air is lost to the surrounding atmosphere. If the return air damper is open and the outside damper closed, then all the air is recirculated. Most HVAC systems operate somewhere between these two extremes.

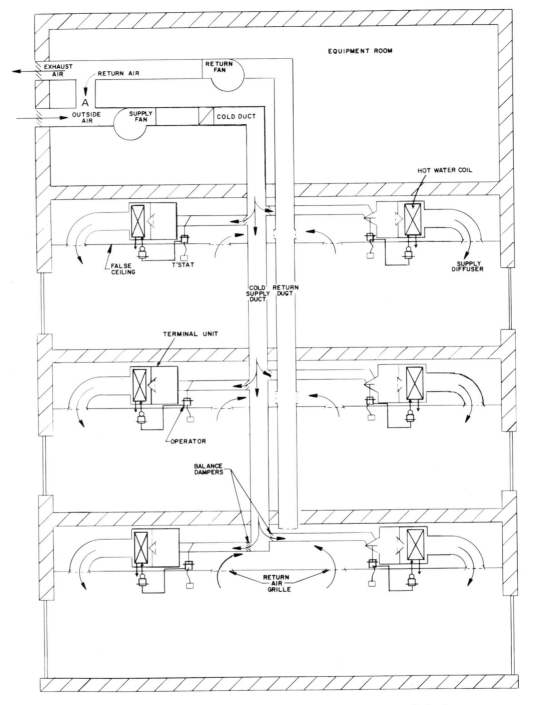

Figure 6–2 Dual-duct air distribution system. (Courtesy of Honeywell, Inc.)

Grilles.   A screen is usually placed upstream from a damper to catch bugs, lint, and debris before they go into the air distribution system. When a grille gets plugged or is blocked (with furniture or stored material), this part of the air distribution system does not work. The authors saw a school where the temperature was consistently too hot for comfort although the control system and other components seemed to be functioning properly. The grilles, however, were completely plugged with lint. When the grilles were cleaned, the problem disappeared.

Fans.   Fans provide the power to move air through the air distribution system. A typical fan has three main parts: a motor, belts or a chain to transmit power from the motor to the fan blades, and the blades with their housing. If any one of these parts fails or is not connected properly, the fan will not move air, and this part of the air conditioning system will not work. The maintenance of each of these components is discussed in Chapter 10.

Heat Exchange Surfaces.   Fans push air, but the air usually must be cooled or heated in order to be useful. This heating or cooling takes place when air is forced around a coil or finned surface containing hot or cold fluid. If these heat exchange surfaces are fouled (with dirt or grease, for example), heat exchange will be inefficient, and more heat or cooling energy will have to be used in order to heat or cool air to the desired temperature.

Ductwork.   Ductwork directs and conducts air from the heat exchange surfaces to the rooms where the hot or cold air is desired, and it conducts the exhaust air from these rooms back to the mixing plenum and to the outside. This function is impaired if the ducts leak or if loose insulation or other obstructions slow the airflow within ducts.

Control System.   An HVAC control system transforms operating policies into the air temperatures and ventilation volumes in the working environment. The policies take the form of room temperatures and ventilation air volumes to be maintained by the HVAC system; the control system has the task of regulating the HVAC system so that these conditions are met as nearly as possible.

The control system accomplishes its function through a system of sensors, actuators, and communication links. The sensors send appropriate electrical or pneumatic signals when some temperature, pressure, or humidity threshold has been crossed. These signals are sent through a communication network, generally either electrical or pneumatic (although multiplexing through electrical supply lines is sometimes used, as is radio transmission). The signals are translated into pneumatic or electrical signals that are then used to open or close dampers, to regulate fans, and to initiate or stop the source of heating or cooling. Dampers can be opened or closed to regulate the amount of incoming air, the amount of exhaust air that is mixed with fresh air, and the amount of air that is introduced into an area. Fans can be turned on or off to increase air coming into a room or being exhausted from it, or their speed can be regulated so that the amount of air coming into a room is no more than is needed for proper ventilation and temperature control. Boilers or chillers can be turned on to provide heat or cooling for the heat exchangers described. Boiler controls are discussed in more detail in Chapter 8.

6.2.2 Other HVAC System Types

Section 6.2.1 described the working of a dual-duct system. Other common HVAC system types are variable air volume (VAV), terminal reheat, fan coil, unit ventilator, induction, steam, and hot water systems. These systems are described briefly below because someone performing an energy audit will invariably run into each of them at some time. For more detail, see References 6 and 7.

6.2.2.1 VAV systems.   In a variable air volume system, tempered (heated or cooled) air is forced into a room at a rate dependent on the amount of heating or cooling desired. If less heating is desired, less hot air is blown into the room. Volume controls are usually located at the duct opening where the air enters the room. The advantage of this system is that only the amount of air needed is used, and, since power requirements vary as the cube of air volume moved, less air volume means less electrical consumption. Some of the disadvantages of this system include the complexity and difficulty of maintenance of the controls and the necessity to use and control high-velocity airstreams.

6.2.2.2 Terminal reheat systems.   Cold air passes through a single duct and enters a zone after passing through a reheat coil which heats the air in accordance with the control needs. Where the source of heat for the reheat coil is a boiler, a common design fault is to have pumps continuously running water from the boiler through the coil system. This uses electricity constantly for the pumps but avoids thermal shock that might occur if cold water were injected into a warm boiler. A better alternative is to have a mixing valve installed at the boiler so that pumps can be shut down when heating is not needed.

6.2.2.3 Fan coil systems.   The dual-duct system and the two systems just described provide ventilation as well as heating and cooling. The function of the fan coil system, however, is to provide heat or cooling only. In such a system, a fan moves room air across heating or cooling coils and back into the room. Control is provided by varying the amount of heating or cooling fluid and/or the fan speed [7, p. 13.7].

6.2.2.4 Unit ventilator.   In this system, air is brought in directly from the outside and heated or cooled as it enters the room. A *swamp cooler* falls into this category as do many of the individual room units in motels. The advantage of this system is that each room can be individually and easily controlled; the disadvantages are that installation costs are high and the temperatures are usually not controlled so as to minimize energy consumption.

6.2.2.5 Induction units.   In these units, high-pressure air flowing through nozzles induces more airflow from the room. This secondary air flows over heating or cooling coils and back into the room. This system provides both ventilation and heating or cooling at relatively low capital and energy costs. It also gives good local control of temperature. Its disadvantages are that its controls are complex and that each unit must be maintained regularly to keep it free of lint and dust.

6.2.2.6 Steam units.

In these systems, heat comes from steam that condenses in radiators and is transferred by either fans or natural convection. The condensate is then returned to the boiler where the steam was generated. The advantages of such systems include low initial and maintenance expenses for a multiroom installation; disadvantages may include the necessity for operating a boiler when only a small part of the boiler design capacity is needed—when, for example, one or two rooms need heat and when the boiler was designed to meet the needs of an entire building. These systems are discussed in detail in Chapter 13 of Reference 7.

6.2.2.7 Water systems.

Water systems range from complex high-temperature units to the more familiar two-pipe units found in many old apartment buildings. In a typical water system, hot or cold water is pumped through coils and heats or cools air that is drawn around the coil by natural convection or fans. In a two-pipe system, water enters the radiator through one pipe and leaves by another. In this system, complex valving is necessary to be able to change the system from heating to cooling, and the system operators must be skilled. In a four-pipe system, two pipes take and remove hot water and two take and remove chilled water, with the relative amounts of each depending on the amount of heating or cooling desired. The four-pipe system involves more plumbing than the two-pipe system but avoids the necessity for changing from hot to cold water throughout the system.

The main advantage of water systems is that they move a large amount of heating and cooling energy in return for a small amount of pumping energy; the amount of energy distribution energy per unit of heating or cooling is significantly less than that of an air system. The main disadvantage is the large amount of plumbing involved. Piping is expensive to buy and to install, and leaks in piping can cause far more expensive consequences than leaks in air-duct systems.

6.3 IMPROVING THE OPERATION OF THE HVAC SYSTEM

The objective of learning the operation of an HVAC system is the more efficient management of that system. This management can be improved by careful attention to the following operating rules.

Operating Rule 1. Heat to the lowest temperature possible, and cool to the highest temperature possible. In one application of this rule, the hot and cold air temperatures on the hot and cold sides of a dual-duct system are set so that one zone is receiving only hot air and one zone is receiving only cold air. The hot air temperature is thus set so that the system meets the heating needs of the coldest room and the cooling needs of the warmest zone (and automatically meets the temperature needs of all the other zones). Another application of the rule would be to avoid heating warehouses unless they contain people or materials sensitive to heat or to cold.

Operating Rule 2.   Avoid heating or cooling when heating or cooling is not needed. For example, heating or cooling for people is not needed when people are not in a building at times such as weekends or at night. At those times, a building temperature can be allowed to drift with the only constraint being the safe temperature of building components or other material contained within the building.

Operating Rule 3.   Learn how your control system is supposed to work and maintain it. A consistent problem that plagues buildings is a control system that does not work. Because a control system is not functioning as intended, return air dampers are blocked open, and the HVAC system heats or cools all the air used for ventilation. People don't understand the way a two-thermostat system is supposed to work, and they turn the wrong thermostat, causing heating at night but none when the heating is needed.

Operating Rule 4.   Adjust the ventilation system so that the minimum amount of ventilation air is being used by altering the control system settings or by changing pulleys on fans or their drive motors. One very useful relationship is

$$\frac{\text{hp}_A}{\text{hp}_B} = \left(\frac{\text{cfm}_A}{\text{cfm}_B} \right)^3 \tag{6-6}$$

This is the cube law for fan horsepower [8]. It is useful in calculating the energy consumption to be saved from reducing the ventilation requirements. If, for example, the amount of ventilation air can be reduced by one third, then the amount of electrical energy supplied can be reduced to $(\frac{2}{3})^3$, or $\frac{8}{27}$, about 30% of its initial rate. Since 1 hp is equivalent to .746 kw, the power savings achieved by a reduction of H hp is given by

$$\text{power saved (kw)} = \frac{H \times .746}{\text{eff}} \tag{6-7}$$

where eff is the motor efficiency, usually between .50 and .80. If this motor is running constantly, the action of reducing the fan horsepower reduces both the demand and the energy part of the electric bill; otherwise it may affect the demand but must be multiplied by the hours of use to determine the amount and cost of energy saved.

Operating Rule 5.   If you don't need it on, turn it off. Unnecessary ventilation costs money. Find out when the ventilation is needed and arrange to have it running only at those times.

6.4 SUMMARY

In this chapter, we have presented HVAC systems in a way that provides an understanding of their functions and components. Once the functions are understood, the energy manager can find ways to reduce the heating and cooling loads and to thus reduce this component of energy costs. When the manager also understands how each of the HVAC system components works, he or she is then prepared to improve the operation of the physical system. With the use of the operating rules presented, and with additional rules discovered by the manager, this understanding can then be translated into improved operating policies for both HVAC equipment and for the people affected by the HVAC system.

REFERENCES

1. *ASHRAE Handbook and Product Directory: 1981 Fundamentals,* American Society of Heating, Refrigerating, and Air Conditioning Engineers, Atlanta, Ga., 1981.

2. Donald Rapp, *Solar Energy,* Prentice-Hall, Englewood Cliffs, N.J., 1981.

3. W. J. Kennedy, Jr., and Richard E. Turley, "Scheduling and the Energy Economics of Wall Design," in *Proceedings of the 1981 Spring Annual Conference & World Productivity Conference, Detroit, Michigan,* Institute of Industrial Engineers, Atlanta, Ga., 1981.

4. Larry G. Berglund and Richard R. Gonzalez, "Human Response to Temperature Drift," *ASHRAE Journal,* Aug. 1978, pp. 38–41.

5. L. G. Berglund, and R. R. Gonzalez, "Application of Acceptable Temperature Drifts to Built Environments as a Mode of Energy Conservation," *ASHRAE Transactions,* Vol. 84, Part 1, 1978.

6. National Electric Contractors Association and National Electrical Manufacturers Association, *Total Energy Management: A Practical Handbook on Energy Conservation and Management,* National Electrical Contractors Association, Washington, D.C., 1976.

7. *ASHRAE Handbook and Product Directory: 1980 Systems,* American Society of Heating, Refrigerating, and Air Conditioning Engineers, Atlanta, Ga., 1980.

8. *ASHRAE Handbook and Product Directory: 1979 Equipment,* American Society of Heating, Refrigerating, and Air Conditioning Engineers, Atlanta, Ga., 1979.

BIBLIOGRAPHY

Industrial Ventilation, Committee on Industrial Ventilation, American Conference of Governmental Industrial Hygienists, Lansing MI, 1979.

Variable Air Volume Systems Manual, 1980 ed., Honeywell, Inc., Minneapolis, 1980.

QUESTIONS

6.1. You are performing an energy audit in the winter on a school with a dual-duct HVAC system. In one room, you note that the temperature is 55°F although the thermostat is set at 70°. Explain how this discrepancy could be caused by (a) dampers, (b) grilles, (c) fans, (d) filters or ductwork, (e) the boiler, or (f) the control system.

6.2. What factors other than those discussed in the text would need to be considered in determining the heating and cooling requirements for a building?

PROBLEMS

6.1. Estimate the total heating load caused by a work force of 22 people including 6 overhead personnel, primarily sitting during the day; 4 maintenance personnel and supervisors; and 12 people doing heavy labor. Assume that everyone works the same 8-h day.

6.2. If the HVAC system that removes the heat in Problem 6.1 has an efficiency of 60% and runs continuously, how many kw will this heating load contribute to the electrical peak if the peak usually occurs during the working day? Assume that the motors in the HVAC system are outside the system and do not contribute to the cooling load.

6.3. Answer Problem 6.2 under the assumption that 8 of the 12 people doing heavy labor and 2 foremen-maintenance personnel come to work when the others are leaving and that 3000 w of extra lighting are required for the night shift.

6.4. A heated building has six window panes gone, each 8 × 10 in., on the windward side. The wind speed has been measured at 900 ft/min, and the location has 6000 heating degree days/year. The mean humidity ratio is .010 lb of moisture per pound of dry air. (a) Calculate the total number of Btu lost through these windows per year. (b) If the heat is supplied by a boiler and heat generation and transmission efficiency is 60%, estimate the cost of leaving the windows broken if gas costs $1.20/dth.

6.5. You have measured the ventilation in a large truck bay and have found that you are using 12,000 cfm. An analysis shows that only 8000 cfm are required. Measurements at the fans give the total electrical consumption of the HVAC system as 16.0 kw at current cfm rates. You are currently ventilating this area 16 h each day, 250 days each year, including the times of peak electrical usage. Your monthly electric rates are $.045/kwh and $12.00/kw of demand. Assuming that your power factor is at least 90% and that your marginal electrical costs are at the least expensive rates, what is the amount of annual savings that you can expect by the proposed reduction in ventilation rates?

6.6. After implementing the improvements suggested in Problem 6.5, you decide to analyze the value of having the second shift come in just as the first shift is leaving, thereby reducing the amount of time that ventilation is needed by 1 h each day. How much annual savings do you expect this measure to achieve?

....

Chapter 7

Combustion Processes
and the Use
of Industrial Wastes

7.0 INTRODUCTION

Any person involved in energy management needs to know what operating parameters of a boiler are important, since boilers often consume 90% or more of the fuel used in a facility. One critical boiler parameter is the amount of combustion air to be mixed with the fuel. The proper amount of air depends on the fuel type, the amount of energy demanded from the boiler per hour, the boiler type, and other considerations, many of which are related to the chemistry of the combustion process. Combustion creates products, some of which are subject to governmental regulation, and the amount and type of these products also depend on the chemical reactions of combustion. A major function of this chapter is, then, to describe the chemistry of combustion in a boiler. Flue gas analysis is also discussed here.

Having laid the foundation in basic combustion chemistry, in this chapter we also describe the chemistry and combustion problems associated with alternative fuels, including industrial wastes. Formerly the use of industrial wastes for fuel was ignored because of obvious technical problems, but the increasing costs of waste disposal and of energy have caused a reexamination of alternatives with sometimes surprising results. As an example, one pulp mill in the South initially estimated that converting existing boilers from coal to pulp mill liquor would cost $1 million in capital and installation cost, and the project was dropped. Government regulations forced the mill to consider alternatives to dumping the liquor into a local river, and the mill found that using it as an alternative fuel would save $1 million per year!

A third objective of this chapter is to illuminate some of the other economic and technical problems to be overcome in changing fuels and to give the reader a set of procedures to use in examining alternative fuel proposals. The potential and the problems inherent in some of the new boiler types are discussed.

7.1 COMBUSTION CHEMISTRY

7.1.1 Basic Reactions and Quantities

When fuels burn, oxygen combines with chemical components of the fuel and releases energy in the form of heat. Since there are many components in fuel, there are many chemical reactions. Some of them are given by*

$$C + O_2 \rightarrow CO_2 + 14{,}093 \text{ Btu/lb} \qquad (7-1)$$

$$2C + O_2 \rightarrow 2CO + 3960 \text{ Btu/lb} \qquad (7-2)$$

$$2H_2 + O_2 \rightarrow 2H_2O + 61{,}100^\dagger \text{ Btu/lb} \qquad (7-3)$$

$$CH_4 + 2O_2 \rightarrow 2H_2O + CO_2 + 23{,}879 \text{ Btu/lb} \qquad (7-4)$$
$$\text{(methane reaction)}$$

$$S + O_2 \rightarrow SO_2 + 3980 \text{ Btu/lb} \qquad (7-5)$$

To understand the implications of these reactions for boiler operation, it is necessary to know the constituents of any boiler fuel by weight percentage and to understand the weight balances implied by the reactions. It is also necessary to know which of the reactions are favored at a given temperature and to know what temperatures are to be used in the computations. With this information, it is possible to calculate the optimum quantity of air needed for combustion of each ton of fuel consumed and to determine the cost of operating at air quantities that exceed this optimum.

To start, consider Reaction 7–4. This represents a balance in molecules; i.e., one molecule of methane combines with two molecules of oxygen to yield two molecules of water and one molecule of carbon dioxide. Under the assumption of Avogadro's law that equal numbers of molecules of dissimilar gases occupy the same amounts of volume, Reaction 7–4 becomes a statement about volumes; i.e., 1 ft^3 of methane combines with 1ft^3 of oxygen to yield 2 ft^3 of water vapor and 1 ft^3 of carbon dioxide. To convert Reaction 7–4 into a statement about weights requires a knowledge of the atomic weights of each component. The atomic weights of carbon,

---

*The detailed chemistry of boiler combustion is very complicated and includes many reactions other than these, for example, those that result in the formation of nitrogen and sulfur oxides. The reactions presented here are, however, adequate for the purposes of energy management and give good quantitative results. For more detail, see Reference 1.

†These figures assume that all of the water leaves as water vapor. The actual value is up to 10% less than this, depending on the amount of water vapor that condenses. Since this amount is usually not known, the authors have chosen to use the higher figure.

hydrogen, and oxygen are, respectively, 12.005, 1.008, and 16.000. With these data, Reaction 7-4 becomes a weight balance:

$$\text{methane weight} + \text{oxygen weight} = \text{water vapor weight} + CO_2 \text{ weight} \quad (7\text{-}6)$$

To get specific values for Equation 7-6, some volume must be assumed. The number of pounds of a gas equal to its molecular weight occupies 359 ft³ (the *molal volume*) where the volume is computed at standard temperature and pressure (32°F and 14.7 psi absolute). The quantities in Equation 7-6 come from relative volumes, given by Reactions 7-1 through 7-5, and weights, calculated from the molecular weights of the constituents in those reactions. With these, Equation 7-6 becomes a statement that can be applied to any volume, once the weight of one compound in that volume has been determined. For a molal volume, the components of Equation 7-6 become

$$
\begin{aligned}
\text{methane weight} &= \text{weight of carbon} + \text{weight of hydrogen} \\
&= 12.005 \text{ lb} + 4 \text{ (atoms of hydrogen per molecule of methane)} \\
&\quad \times 1.008 \text{ lb} \\
&= 16.037 \text{ lb} \\
\text{oxygen weight} &= 2 \text{ (proportionality constant)} \\
&\quad \times 2 \text{ (atoms of atomic oxygen per atom of molecular oxygen)} \\
&\quad \times 16.000 \text{ lb} \\
&= 64 \text{ lb} \\
H_2O \text{ weight} &= 2 \times (2 \times 1.008 + 16) \\
&= 36.032 \text{ lb} \\
CO_2 \text{ weight} &= 12.005 + 2 \times 16.000 \\
&= 44.005 \text{ lb}
\end{aligned}
$$

These equations yield the following mass balance:

$$16.037 \text{ lb methane} + 64 \text{ lb oxygen} \rightarrow 36.032 \text{ lb water vapor} \quad (7\text{-}7)$$
$$+ \ 44.005 \text{ lb } CO_2$$

These relationships assume complete combustion of all carbon in the fuel, i.e., that the fuel is completely converted into water and CO_2. Such complete combustion occurs only with perfect mixing of air and fuel or in the presence of excess combustion air. Perfect mixing would ensure that every molecule of fuel is surrounded by the right amount of air for combustion to take place and is rarely, if ever, attained. Since the products of incomplete combustion are less desirable than the heating associated with excess air, excess air is preferred, in amounts shown in Table 7-1. Usually, the objective of combustion air control is to introduce enough air to keep the concentration of CO below 400 ppm (for gas fuels) or the smoke color clear or light brown (for liquid fuels or coal). A safety factor is added to this air requirement to keep the com-

TABLE 7.1 RECOMMENDED EXCESS AIR
FOR VARIOUS FUELS

Fuels		% excess air
Solid	Coal (pulverized)	15–30
	Coke	20–40
	Wood	25–50
	Bagasse	25–45
Liquid	Oil	3–15
Gaseous	Natural gas	5–10
	Refinery gas	8–15
	Blast-furnace gas	15–25
	Coke-oven gas	5–10

Source: Courtesy of Combustion Engineering, Inc.

bustion in an oxygen-rich environment. With 100% excess oxygen supplied by combustion air, the equation for combustion becomes

$$16.037 \text{ lb methane } + \text{ 128 lb oxygen}$$
$$\rightarrow 36.032 \text{ lb water vapor } + \text{ 44.005 lb CO}_2 + \text{ 64 lb excess oxygen} \qquad (7\text{-}8)$$
$$+ \text{ 382,947 Btu}$$

To translate this into an equality related to combustion *air,* it is necessary to know the number of pounds of oxygen per pound of air. Since oxygen makes up approximately 23.15% of the weight of air, Equation 7-8 gives

$$16.037 \text{ lb methane } + \text{ 552.916 lb air}$$
$$\rightarrow 36.032 \text{ lb water vapor } + \text{ 44.005 lb CO}_2 + \text{ 276.458 lb excess air} \qquad (7\text{-}9)$$
$$+ \text{ 382,947 Btu}$$

Another useful relationship is developed by using the density of air, .07655 lb/ft³ for dry air at sea level, to get

$$16.037 \text{ lb CH}_4 + \text{ 7222.9 ft}^3 \text{ dry air}$$
$$\rightarrow 36.032 \text{ lb water vapor } + \text{ 44.005 lb CO}_2 + \text{ 3611.5 ft}^3 \text{ dry air} \qquad (7\text{-}10)$$
$$+ \text{ 382,947 Btu}$$

The preceding analysis gives a method for determining the amount of dry air and the amount and type of chemical products of combustion for a given fuel constituent, in this case methane. Table 7-2 gives this information for many substances and is often used to estimate the heat liberated and the excess air needed when a particular gas of known composition is contemplated as a fuel.

Example

Consider a gas (typical of a natural gas being supplied to users) with the analysis given in Figure 7-1. The amount of air needed for combustion per cubic foot of gas, assuming perfect mixing, is given by .864 × 9.528 + .084 × 16.675 + .015 × 23.821 + .011 × 30.967, or 10.33 ft³ of air per cubic foot of gas. Assuming that 12% excess air is needed to

TABLE 7.2 COMBUSTION CONSTANTS

Column groups: **Heat of Combustion[c]** = (Btu per Cu Ft: Gross, Net[d]; Btu per Lb: Gross, Net[d]). **Cu Ft per Cu Ft of Combustible** = (Required for Combustion: O₂, N₂, Air; Flue Products: CO₂, H₂O, N₂). **Lb per Lb of Combustible** = (Required for Combustion: O₂, N₂, Air; Flue Products: CO₂, H₂O, N₂).

No.	Substance	Formula	Mol. Weight[a]	Lb per Cu Ft[b]	Cu Ft per Lb[b]	Sp Gr Air=1.000[b]	Btu/CuFt Gross	Btu/CuFt Net[d]	Btu/Lb Gross	Btu/Lb Net[d]	Req O_2 (cuft)	Req N_2 (cuft)	Req Air (cuft)	Flue CO_2 (cuft)	Flue H_2O (cuft)	Flue N_2 (cuft)	Req O_2 (lb)	Req N_2 (lb)	Req Air (lb)	Flue CO_2 (lb)	Flue H_2O (lb)	Flue N_2 (lb)	Exp. Error % + or −
1	Carbon	C	12.01	—	—	—	—	—	14,093[g]	14,093[g]	—	—	—	—	—	—	2.664	8.863	11.527	3.664	—	8.863	0.012
2	Hydrogen	H_2	2.016	0.005327	187.723	0.06959	325.0	275.0[d]	61,100	51,623	0.5	1.882	2.382	—	1.0	1.882	7.937	26.407	34.344	—	8.937	26.407	0.015
3	Oxygen	O_2	32.000	0.08461	11.819	1.1053	—	—	—	—	—	—	—	—	—	—	—	—	—	—	—	—	—
4	Nitrogen (atm)	N_2	28.016	0.07439[e]	13.443[e]	0.9718[e]	—	—	—	—	—	—	—	—	—	—	—	—	—	—	—	—	—
5	Carbon monoxide	CO	28.01	0.07404	13.506	0.9672	321.8	321.8	4,347	4,347	0.5	1.882	2.382	1.0	—	1.882	0.571	1.900	2.471	1.571	—	1.900	0.045
6	Carbon dioxide	CO_2	44.01	0.1170	8.548	1.5282	—	—	—	—	—	—	—	—	—	—	—	—	—	—	—	—	—
Paraffin series C_nH_{2n+2}																							
7	Methane	CH_4	16.041	0.04243	23.565	0.5543	1013.2	913.1	23,879	21,520	2.0	7.528	9.528	1.0	2.0	7.528	3.990	13.275	17.265	2.744	2.246	13.275	0.033
8	Ethane	C_2H_6	30.067	0.08029[a]	12.455[e]	1.04882[e]	1792	1641	22,320	20,432	3.5	13.175	16.675	2.0	3.0	13.175	3.725	12.394	16.119	2.927	1.798	12.394	0.030
9	Propane	C_3H_8	44.092	0.1196[e]	8.365[e]	1.5617[e]	2590	2385	21,661	19,944	5.0	18.821	23.821	3.0	4.0	18.821	3.629	12.074	15.703	2.994	1.634	12.074	0.023
10	n-Butane	C_4H_{10}	58.118	0.1582[e]	6.321[e]	2.06654[e]	3370	3113	21,308	19,680	6.5	24.467	30.967	4.0	5.0	24.467	3.579	11.908	15.487	3.029	1.550	11.908	0.022
11	Isobutane	C_4H_{10}	58.118	0.1582[e]	6.321[e]	2.06654[e]	3363	3105	21,257	19,629	6.5	24.467	30.967	4.0	5.0	24.467	3.579	11.908	15.487	3.029	1.550	11.908	0.019
12	n-Pentane	C_5H_{12}	72.144	0.1904[e]	5.252[e]	2.4872[e]	4016	3709	21,091	19,517	8.0	30.114	38.114	5.0	6.0	30.114	3.548	11.805	15.353	3.050	1.498	11.805	0.025
13	Isopentane	C_5H_{12}	72.144	0.1904[e]	5.252[e]	2.4872[e]	4008	3716	21,052	19,478	8.0	30.114	38.114	5.0	6.0	30.114	3.548	11.805	15.353	3.050	1.498	11.805	0.071
14	Neopentane	C_5H_{12}	72.144	0.1904[e]	5.252[e]	2.4872[e]	3993	3693	20,970	19,396	8.0	30.114	38.114	5.0	6.0	30.114	3.548	11.805	15.353	3.050	1.498	11.805	0.11
15	n-Hexane	C_6H_{14}	86.169	0.2274[e]	4.398[e]	2.9704[e]	4762	4412	20,940	19,403	9.5	35.760	45.260	6.0	7.0	35.760	3.528	11.738	15.266	3.064	1.464	11.738	0.05
Olefin series C_nH_{2n}																							
16	Ethylene	C_2H_4	28.051	0.07456	13.412	0.9740	1613.8	1513.2	21,644	20,295	3.0	11.293	14.293	2.0	2.0	11.293	3.422	11.385	14.807	3.138	1.285	11.385	0.021
17	Propylene	C_3H_6	42.077	0.1110[e]	9.007[e]	1.4504[e]	2336	2186	21,041	19,691	4.5	16.939	21.439	3.0	3.0	16.939	3.422	11.385	14.807	3.138	1.285	11.385	0.031
18	n-Butene (Butylene)	C_4H_8	56.102	0.1480[e]	6.756[e]	1.9336[e]	3084	2885	20,840	19,496	6.0	22.585	28.585	4.0	4.0	22.585	3.422	11.385	14.807	3.138	1.285	11.385	0.031
19	Isobutene	C_4H_8	56.102	0.1480[e]	6.756[e]	1.9336[e]	3068	2869	20,730	19,382	6.0	22.585	28.585	4.0	4.0	22.585	3.422	11.385	14.807	3.138	1.285	11.385	0.031
20	n-Pentene	C_5H_{10}	70.128	0.1852[e]	5.400[e]	2.4190[e]	3836	3586	20,712	19,363	7.5	28.232	35.732	5.0	5.0	28.232	3.422	11.385	14.807	3.138	1.285	11.385	0.037
Aromatic series C_nH_{2n-6}																							
21	Benzene	C_6H_6	78.107	0.2060[e]	4.852[e]	2.6920[e]	3751	3601	18,210	17,480	7.5	28.232	35.732	6.0	3.0	28.232	3.073	10.224	13.297	3.381	0.692	10.224	0.12
22	Toluene	C_7H_8	92.132	0.2431[e]	4.113[e]	3.1760[e]	4484	4284	18,440	17,620	9.0	33.878	42.878	7.0	4.0	33.878	3.126	10.401	13.527	3.344	0.782	10.401	0.21
23	Xylene	C_8H_{10}	106.158	0.2803[e]	3.567[e]	3.6618[e]	5230	4980	18,650	17,760	10.5	39.524	50.024	8.0	5.0	39.524	3.165	10.530	13.695	3.317	0.849	10.530	0.36
Miscellaneous gases																							
24	Acetylene	C_2H_2	26.036	0.06971	14.344	0.9107	1499	1448	21,500	20,776	2.5	9.411	11.911	2.0	1.0	9.411	3.073	10.224	13.297	3.381	0.692	10.224	0.16
25	Naphthalene	$C_{10}H_8$	128.162	0.3384[e]	2.955[e]	4.4208[e]	5854[f]	5654[f]	17,298	16,708[f]	12.0	45.170	57.170	10.0	4.0	45.170	2.996	9.968	12.964	3.434	0.562	9.968	—[f]
26	Methyl alcohol	CH_3OH	32.041	0.0846[e]	11.820[e]	1.1052[e]	867.9	768.0	10,259	9,078	1.5	5.646	7.146	1.0	2.0	5.646	1.498	4.984	6.482	1.374	1.125	4.984	0.027
27	Ethyl alcohol	C_2H_5OH	46.067	0.1216[e]	8.221[e]	1.5890[e]	1600.3	1450.5	13,161	11,929	3.0	11.293	14.293	2.0	3.0	11.293	2.084	6.934	9.018	1.922	1.170	6.934	0.030
28	Ammonia	NH_3	17.031	0.0456[e]	21.914[e]	0.5961[e]	441.1	365.1	9,668	8,001	0.75	2.823	3.573	—	1.5	3.323	1.409	4.688	6.097	—	1.587	5.511	0.088
29	Sulfur	S	32.06	—	—	—	—	—	3,983	3,983	—	—	—	—	—	—	0.998	3.287	4.285	1.998 (SO_2)	—	3.287	0.071
30	Hydrogen sulfide	H_2S	34.076	0.09109[e]	10.979[e]	1.1898[e]	647	596	7,100	6,545	1.5	5.646	7.146	1.0 (SO_2)	1.0	5.646	1.409	4.688	6.097	1.880 (SO_2)	0.529	4.688	0.30
31	Sulfur dioxide	SO_2	64.06	0.1733	5.770	2.264	—	—	—	—	—	—	—	—	—	—	—	—	—	—	—	—	—
32	Water vapor	H_2O	18.016	0.04758[e]	21.017[e]	0.6215[e]	—	—	—	—	—	—	—	—	—	—	—	—	—	—	—	—	—
33	Air		28.9	0.07655	13.063	1.0000	—	—	—	—	—	—	—	—	—	—	—	—	—	—	—	—	—

All gas volumes corrected to 60F and 30 in. Hg dry. For gases saturated with water at 60F, 1.73% of the Btu value must be deducted.

[a] Calculated from atomic weights given in "Journal of the American Chemical Society", February 1937.

[b] Densities calculated from values given in grams per liter at 0C and normal pressure in the International Critical Tables allowing for the known deviations from the gas laws. Where the coefficient of expansion was not available, the assumed value was taken as 0.0037 per °C. Compare this with 0.003662 which is the coefficient for a perfect gas. Where no densities were available the volume of the mol was taken as 22.4115 liters.

[c] Converted to mean Btu per lb (1/180 of the heat per lb of water from 32F to 212F) from data by Frederick D. Rossini; National Bureau of Standards, letter of April 10, 1937, except as noted.

[d] Deduction from gross to net heating value determined by deducting 18,919 Btu per pound mol of water in the products of combustion. Osborne, Stimson, and Ginnings, "Mechanical Engineering", p. 163, March 1935, and Osborne, Stimson, and Fiock, National Bureau of Standards Research Paper 209.

[e] Denotes that either the density or the coefficient of expansion has been assumed. Some of the materials cannot exist as gases at 60F and 30 in. Hg pressure, in which case the values are theoretical ones given for ease of calculation of gas problems. Under the actual concentrations in which these materials are present their partial pressure is low enough to keep them as gases.

[f] From Third Edition of "Combustion."

[g] National Bureau of Standards, RP 1141.

Reprinted from "Fuel Flue Gases", 1941 Edition, courtesy of American Gas Association.

Constituent	Percent by volume	Btu/ft³ of mixture
CH_4	86.4	788.9 (= .864 × 913.2)
C_2H_6	8.4	150.1
C_3H_8	1.5	38.9
C_4H_{10}	1.1	37.1
N_2	.5	—
CO_2	2.1	—

Total Btu/ft³: 1015.0

Figure 7-1 Composition of a typical fuel gas, by volume.

assure complete combustion, 1.24 ft³ of air in addition to the preceding are required, for a total of 11.57 ft³ of air per cubic foot of gas.

These methods can also be used to estimate the number of cubic feet per minute of combustion air that must be supplied to provide a given Btu output. (Note that in designing a system, the components must be sized for more capacity than the average to assure that delivery capacity is always available for the required amount of combustion air.) If a boiler with the preceding fuel is to deliver 2 million Btu/h and the boiler efficiency is .80, then the amount of air needed per minute is

$$\frac{2,000,000\ \text{Btu/h}}{.8} \times \frac{1\ \text{h/60 min}}{1015\ \text{Btu/ft}^3\ \text{gas}} \times 11.6\ \text{ft}^3\ \text{air/ft}^3\ \text{gas} = \tag{7-11}$$
$$476\ \text{cfm (ft}^3\text{/min)}$$

Table 7-2 can also be used to calculate the flue gas composition. The calculations are shown in Figure 7-2 for this example under assumed conditions of 12% ex-

Constituent	ft³/ft³ fuel gas	Products: ft³/ft³ fuel gas			
		CO_2	N_2	O_2 (excess)	H_2O
Excess air (preceeding example)	1.24		.980	.260	
CH_4	.864	.864		6.504	1.728
C_2H_6	.84	.168		1.107	.252
C_3H_8	.015	.045		.282	.060
C_4H_{10}	.011	.044		.269	.055
N_2	.005			.005	
CO_2	.021	.021			
Total		1.142	9.147	.260	2.095
Fraction of total volume		.090	.723	.021	.166

Figure 7-2 Volumetric analysis of flue gas for a gas fuel.

cess air. Since air is approximately 76.85% N_2 and 23.15% O_2 by weight [2], excess air of .12 lb/lb air needed for combustion translates into .0922 lb N_2 and .0278 lb O_2.

If an analysis by weight is desired, two methods are available. The preceding calculations can be performed using weight as a basis from Table 7-1, calculating the weight of combustion air needed for each constituent, determining the weight of excess air, calculating the number of pounds of N_2 and O_2 represented by this weight, and tabulating the results. Or the total volume fractions from the preceding calculations can be converted directly into weight calculations. The latter method works as shown in Figure 7-3.

Constituent	ft^3/ft^3 flue gas	Density (lb/ft^3)	lb/ft^3 flue gas	Weight %
CO_2	.090	.1170	.01053	14.2
N_2	.723	.07439	.05378	72.7
O_2	.021	.08461	.00178	2.4
H_2O	.166	.04758	.00790	10.7
Total weight, lb/ft^3:			.07399	100.0

Figure 7–3   Weight analysis of flue gas for a gas fuel.

The analysis methods used for coal and other nongas fuels differ from the gas analysis already shown. For these fuels, a *proximate* analysis gives the percentage by weight of moisture, volatile matter, fixed carbon, and ash. A proximate analysis is usually provided at the time coal is purchased and may be part of the basis for the price paid for the coal. An *ultimate* analysis is the determination, from a dried sample, of carbon, hydrogen, sulfur, and nitrogen and an estimate of the amount of oxygen as the difference between the starting weight and the weight of all the other constituents. Typical, though hypothetical, values are shown in Figure 7–4, together with

Proximate analysis		Ultimate analysis	
Component	Weight %	Component	Weight %
Moisture	3.5	Moisture	3.5
Volatile matter	42.4	Carbon	76.2
Fixed carbon	51.3	Hydrogen	6.2
Ash	2.8	Sulfur	1.8
Total: 100.0		Nitrogen	2.6
		Oxygen	6.9
		Ash	2.8
		Total: 100.0	

Figure 7–4   Proximate and ultimate analyses for a typical coal.

Heating value, Btu/lb: 12,700

a typical heating value. These data can be used to calculate the amount of combustion air as shown in Figure 7–5.

The dry air equivalent of this weighs 2.471 lb O_2 × 100 lb dry air/23.15 lb O_2 = 2.471 × 4.32 = 10.675 lb. At 125% total air (25% excess air), 13.343 lb of dry air would be needed per pound of fuel, and the flue gas would contain 13.343 − 10.675 = 2.668 lb of excess air with .25 × 2.471 = .618 lb excess O_2 per pound of fuel. These calculations are useful in analyzing the flue gas.

For the 2 million Btu/h boiler discussed earlier, the combustion air requirement, assuming 25% excess air by weight, is

$$\frac{2,000,000 \text{ Btu/h}}{.80} \times \frac{1 \text{ h/60 min}}{12,700 \text{ Btu/lb}} \times 13.34 \text{ lb air/lb fuel}$$
$$= 43.76 \text{ lb dry air/min} \qquad (7\text{–}12)$$
$$= 571.6 \text{ cfm}$$

	Ultimate analysis: lb/lb fuel as fired	lb O_2/lb combustible element (from Table 7-2)	Required for combustion, lb O_2/lb fuel (perfect air mixing)
C	.762	2.664	2.030
H_2	.062	7.937	.492
S	.018	.998	.018
N_2	.026		——
O_2	.069		——
H_2O	.035		——
Ash	.028		——
Total:	1.000		Total: 2.540
		Less O_2 in fuel:	−.069
			2.471

Figure 7–5 Calculation of combustion air quantities.

The amount of each combustion product in the flue gas can also be calculated from the ultimate analysis using the "flue products" columns of Table 7–2 and assuming that *standard air* (air with .013 lb/lb dry air at 80°F dry bulb; relative humidity, 60% and pressure 14.7 psia) is used. These calculations proceed as shown in Figure 7–6 for the coal described. Dividing the component weights by the total weight (wet or dry, depending on the analysis method used) gives the components by percentage: CO_2, 19.3%; H_2O, 5.3%; SO_2, .25%; N_2, 70.9%; O_2, 4.3% (percentages of wet weight measurement).

7.1.2 Energy Loss Calculations

In Section 7.1.1 we described some of the basic calculations associated with chemical reactions of combustion. A person involved in energy management, however, needs to know more than combustion chemistry. He or she is interested in the aspects of

Analysis: lb/lb fuel as fired		Product	Product/lb constituent					Lb product/lb fuel (assuming 25% excess air)
C	.762	CO_2	3.664	$\times$	.762 =	2.792		2.792 lb
H_2	.062	H_2O	8.94	$\times$	.062 =	.554		
H_2O (in fuel)	.035					.035		
Air (additional H_2O)	13.343		.013	$\times$	13.343 =	.173		
						.762		.762 lb
S	.018	SO_2	.018	$\times$	1.998 =	.036 lb		.036 lb
N_2 (in fuel)	.026					.026		
Air (N_2 component)	13.343		.7685	$\times$	13.343 =	10.254		
						10.280		10.280 lb
O_2 (excess)								.618 lb
			Weight (wet)					14.488 lb
			Weight (dry) = 14.488 − .762 =					13.726 lb

Figure 7-6 Calculation of flue gas composition by weight. (Note that .173 lb extra H_2O is included. The weight balance is 13.343 lb dry air + .173 lb water in air + 1 lb fuel − .028 lb ash = 14.488 lb. This checks with the wet weight of total products.)

combustion that can be controlled so as to make the boiler more efficient, i.e., so as to make it use less fuel for a given output of usable heat energy. This efficiency is governed by the amount of heat that is lost in flue gas, an amount that is dependent on the amount of combustion air provided and that can be computed from the flue gas composition and from the stack gas temperature. In this section we present those computations.

Using excess air leads to complete combustion but causes heat losses to the environment. The heat of combustion raises the temperature of all the combustion products sent up the flue, and the heat loss in these products is a major source of boiler inefficiency. Some of this loss cannot be avoided—it is necessary to remove combustion products, and higher temperatures help the flue to exhaust these gases naturally. The flue also exhausts unburned gases, including excess combustion air, after adding enough heat to raise the temperature of the air from ambient to stack temperature, so excess air causes some heat losses. Without some excess air, however, some of the carbon in the fuel becomes CO rather than CO_2, and since the reaction giving CO yields 3960 Btu/lb rather than the 14,093 Btu/lb associated with CO_2, lack of excess air causes a loss of energy. The following example shows how to quantify these relationships.

To calculate the heat loss through combustion products in the coal example we have given, the percent by volume of each flue gas component is first obtained, either by an Orsat analysis (one standard wet chemical analysis method) or by converting the weight analysis already shown into a volume analysis after dividing the weight of each component by its molecular weight. (Remember that the weight of a given volume of gas divided by its molecular weight gives the number of *lb·mol* of the gas, a value proportional to the volume the gas occupies.) It is also necessary to know the exit temperature of the flue gas, the specific heat of each component (from Figure 7–2) at the average of the ambient and exit temperatures, and the heating value of the fuel.

Using the preceding example, assume that the exit temperature is 500°F, the ambient temperature is 80°F, and the heating value is 12,600 Btu/lb. The relative volumes are calculated as shown in Figure 7–7. Note that 1 lb of this fuel combines with

Component	lb/lb flue gas	/	Molecular weight	=	lb · mol/lb flue gas
CO_2	.193		44		.00439
H_2O	.053		18		.00294
SO_2	.0025		64		.00004
N_2	.709		28		.02532
O_2	.043		32		.00134
				Total:	.03403

Figure 7–7 Energy losses in flue gas, calculation 1.

combustion air to create 14.488 lb of wet flue gas or 13.726 lb of dry gas and that 1 lb of wet gas represents the exhaust equivalent of $12,600/14.488 = 869.7$ Btu. Thus to estimate the percent of heat loss, it is necessary to divide the heat loss associated with 1 lb of flue gas by 869.7 Btu (for this fuel). See Figure 7–8.

To this must be added the latent heat contained in the water vapor that was created as a result of the combustion and in the moisture (if any) contained in the fuel. The amount of water vapor created by combustion is .5 lb·mol/lb H_2 ($= 8.937/18$) in the fuel, .031 lb·mol/lb fuel ($= .062 \times .5$) in the preceding analysis. The molecular weight of water is 18, and hence the amount of water in the incoming fuel moisture is

Component	lb · mol/lb flue gas	X	Specific heat, Btu/lb · mol, °F	X	500-80	=	Btu
H_2O (sensible heat)	.00294		8.2		420		10.1
CO_2	.00439		9.8		420		18.1
SO_2	.00004		10.3		420		.2
N_2	.02532		7.1		420		75.5
O_2	.00134		7.2		420		4.1
						Total:	108.0 Btu

Figure 7–8 Energy losses in flue gas, calculation 2.

.035/18, or .0019 lb·mol/lb fuel. The latent heat contained in the water vapor is (.031 + .0019) lb·mol H_2O/lb fuel × 1040 Btu/lb H_2O × 18 lb H_2O/lb·mol H_2O = 618 Btu/lb fuel, or 618/14.488 = 42.7 Btu/lb flue gas. Thus the total heat lost is 108.0 + 42.7, or 150.7 Btu/lb flue gas. Since the energy used to produce 1 lb of flue gas was 869.7 Btu, the loss in energy is 150.7/869.7, or 17.3%.

The preceding analysis can be used as a basis for evaluating many proposed energy management improvement measures. For example, if waste heat recovery can be used in the flue to reduce the exhaust temperature to 350°F, the heat lost will be 42.7 + (350 − 80)/(500 − 80) × 108.0, or 112.1 Btu/lb of flue gas, for a loss of 112.1/869.7, or 12.9%. This represents an efficiency improvement of 4.4%—a substantial sum when multiplied by the amount paid each year for boiler fuel.

As another example, suppose that the air amount were decreased until the first traces of CO appeared and that this occurred at 10% excess air. Suppose, furthermore, that the stack temperature decreased from 500 to 430°F. From calculations like those following Figure 7–5, the amount of excess O_2 is .10 × 2.471, or .247 lb. The total amount of air needed per pound of fuel is 1.10 × 10.67, or 11.74 lb. Following the logic of Figure 7–6, the amount of product per pound of fuel is as follows: CO_2, 2.792 lb; H_2O, .554 + .035 + 11.74 × .013 = .742 lb; SO_2, .036 lb; N_2, .026 + .7685 × 11.74 = 9.048 lb; and excess O_2, .247 lb. The total weight of flue gas is 12.865 lb (per pound of fuel). Checking the weight balance as before gives the following: fuel weight + weight of added water in air + weight of combustion air − weight of ash (since ash is not removed as a flue product) = 1.000 lb + 11.74 × .013 (= .153) + 11.74 − .028 = 12.865 lb, the total weight of the flue gas products. (This is a very useful check on the calculations.) Furthermore, the energy input per pound of flue gas is 12,600/12.865 = 979.4 Btu. Dividing by 12.865 lb also gives these values for the percentage composition (by weight) of the flue gas: CO_2, 21.7; H_2O, 5.76; SO_2, .28; N_2, 70.3; and O_2, 1.92. At this point, the calculations of Figures 7–7 and 7–8 can be combined as shown in Figure 7–9. This gives a loss in efficiency of 146.8/979.4, or 14.99%.

Component	lb/lb flue gas	/	Molecular weight	×	Specific heat, Btu/lb · mol, °F	×	430-80	=	Btu
CO_2	.217		44		8.1				14.0
H_2O (sensible)	.056		18		9.8				10.7
SO_2	.0028		64		10.1				.2
N_2	.703		28		7.1				62.4
O_2	.019		32		7.2				1.5

Total sensible heat loss: 88.6

Latent heat loss = .056 lb × 1040 Btu/lb: 58.2

Total heat loss: 146.8

Figure 7-9   Combined energy loss calculations for exhaust gas.

Before leaving combustion chemistry, it is important to consider the case of incomplete combustion. For any fuel, incomplete combustion is taking place whenever CO is present in the flue gas. Oxidation of carbon into CO yields 3960 Btu/lb C; oxidation into CO_2 yields 14,093 Btu/lb C. Thus the percentage loss in efficiency due to combustion to CO is given by

$$\text{percent loss} = \frac{10{,}133}{14{,}093} \times 100 \times \frac{\text{no. ppm CO}}{\text{no. ppm CO} + \text{no. ppm CO}_2} \qquad (7\text{-}13)$$

where the concentrations of CO and CO_2 come from the flue gas analysis. If, for example, the concentrations of CO_2 and CO were, respectively, 90,000 and 800 ppm (parts per million, by volume), then the percent loss in efficiency would be

$$\text{percent loss} = \frac{.719 \times 100 \times 800}{90{,}400}$$
$$= .636\%$$

This loss is small, but there are other reasons for avoiding incomplete combustion. Incomplete combustion is associated with CO or smoke or both. CO is poisonous in small quantities, and standards generally keep it below 400 ppm in residences. Smoke fouls the interior of boilers and can also create political problems. The presence of products of incomplete combustion is also inherently unstable chemically, and boiler explosions, though infrequent, can result. Incidentally, it is possible to have both incomplete combustion and excess combustion air if the burner forces combustion products onto a cold surface or in older boiler units.

7.2 BOILER FUELS

The cost and time needed to acquire a knowledge of combustion chemistry are justified by the cost penalties associated with inefficient operation. In the search for more efficient ways to operate, another major step is the examination of the boiler fuel presently used with an eye toward replacement of the fuel by a waste product or by a less expensive or more available fuel. In Section 7.2.1, conventional fuels are discussed together with their general advantages and disadvantages; this discussion forms a base for Section 7.2.2 on less conventional fuels, including various types of industrial waste.

7.2.1 Characteristics of Common Industrial Fuels

The ideal industrial fuel would have the following characteristics: (1) it would *cost* very little per Btu as a raw material. (2) It would be readily *available* in any kind of weather and any international political climate. (3) The on-site *equipment* needed to transport and burn it would be mechanically uncomplicated. (4) *Storage* of the fuel would not present a problem. (5) Its combustion would create no harmful *emissions*.

(6) The *technology* for boilers using this fuel would be well proven. Consider coal, fuel oil, and natural gas in the light of these characteristics.

7.2.1.1 Coal.

The cost of coal as delivered ranges from \$35 to \$70/ton, about \$1.40–\$2.80/million Btu (1984 prices). It is readily available, but it is subject to labor strikes and occasionally to weather-caused transportation problems. At the boiler plant, the coal must be removed from its trucks, railroad cars, or slurry pipeline and transported to the boiler, often with intermediate processing. Burning coal usually involves complicated mechanical equipment such as spreader-stokers or atomizing burners. The technology of this equipment is, however, well developed, and the procedures for its maintenance are well known. Coal storage does not present a major problem, although it (1) requires space, (2) can be subject to spontaneous combustion, (3) requires material handling, and (4) can be subject to freezing. Again, these problems have been seen before, and their solution is well known. Since the Clean Air Act of 1977, the amount of NO_x and SO_x (nitrogen and sulfer oxides) in flue gas emissions have been regulated by law, and hence the amount of these oxides for a given fuel is a matter of some concern. Usual coal-fired boilers operate more efficiently at high (2100–2500°F) temperatures than at lower temperatures. Unfortunately, more NO_x forms at these temperatures, and the residence time in the boiler is not sufficient for it to dissociate into O_2 and N_2. SO_x is also a problem, depending on the sulfur content in the coal being burned. The reduction of NO_x and SO_x to legal levels has necessitated expensive flue gas treatment and has made low-sulfur fuels and low-temperature combustion processes more competitive economically. The technology for mining and burning coal has evolved over many centuries; both the problems and their solutions are well known. Figure 7–4 gives chemical analyses for a typical coal.

7.2.1.2 Fuel oil.

This classification includes everything from No. 1, a distillate, such as the kerosene used for home heating, to No. 6, a heavy residual oil. Prices for No. 6 fluctuate around \$25–\$60/42-gal barrel. The availability of this fuel is not affected by weather, but the world political climate can change the supply dramatically as was proved by the Arab Oil Embargo of 1973. The offloading and storage of fuel oil is usually uncomplicated, but its viscosity increases in cold weather, and, depending on the grade, it may be necessary to provide auxiliary heating before it will flow (see Table 7–3). Methods for solving the viscosity problem are well known. As with coal, NO_x and SO_x emissions can present a problem.

7.2.1.3 Natural gas.

This fuel was discussed earlier in an example of combustion chemistry. Gas has many advantages: It is clean, it has no particulate emissions and its exhaust can therefore be used in gas turbines, it is easy to transport with pipelines, and it is easy to burn. Because of its convenience and because its price was kept low by government regulation for a long time, it has been the preferred fuel for many applications. The decontrol of natural gas has, however, allowed the price of gas to increase and made other sources of energy more attractive than before. This change from gas has been accelerated by the unavailability of gas during one severe winter due to the large increase in demand caused by residential customers. NO_x and SO_x do

TABLE 7.3 CHARACTERISTICS OF FUEL OILS[a]

Grade of fuel oil	Flash point, °C (°F), min.	Pour point, °C (°F), max.	Water and Sediment (vol. %), max.	Carbon residue on 10 % bottoms (%), max.	Ash weight (%), max.	Distillation temp. °C (°F)		
						10% Point, max.	90% point min.	90% point max.
No. 1: Distillate oil intended for vaporizing pot-type burners and other burners requiring this grade	38 or legal (100)	−18[c] (0)	.05	.15	—	215 (420)	—	288 (550)
No. 2: Distillate oil for general purpose heating for use in burners not requiring No. 1	38 or legal (100)	−6[c] (20)	.05	.35	—	—	282[c] (540)	338 (640)
No. 4: Preheating not usually required for handling or burning	55 or legal (130)	−6[c] (20)	.50	—	.10	—	—	—
No. 5 (light): Preheating may be required depending on climate and equipment	55 or legal (130)	—	1.00	—	.10	—	—	—
No. 5 (heavy): Preheating may be required for burning and, in cold climates, may be required for handling	55 or legal (130)	—	1.00	—	.10	—	—	—
No. 6: Preheating required for burning and handling	60 (140)	d	2.00[e]	—	—	—	—	—

[a] It is the intent of these classifications that failure to meet any requirement of a given grade does not automatically place an oil in the next lower grade unless in fact it meets all requirements of the lower grade.

[b] Viscosity values in parentheses are for information only and not necessarily limiting.

[c] Lower or higher pour points may be specified whenever required by conditions of storage or use. When pour point less than −18°C (0°F) is specified, the minimum viscosity for Grade No. 2 shall be 1.8 cST (32.0 SUS) and the minimum 90% point shall be waived.

[d] Where low-sulfur fuel oil is required, Grade 6 fuel oil will be classified as low pour +15°C (60°F) max. or high pour (no max.). Low pour fuel oil should be used unless all tanks and lines are heated.

[e] The amount of water by distillation plus the sediment by extraction shall not exceed 2.00%. The amount of sediment by extraction shall not exceed .50%. A deduction in quantity shall be made for all water and sediment in excess of 1.0%.

not seem to cause the problems with gas as a fuel that they do for coal and fuel oil, and gas remains the fuel of choice for many users. The technology, incidentally, is well known and less complicated than that for coal or fuel oil. The components of a gas fuel include various hydrocarbons; a typical example is the natural gas shown in Figure 7–1.

TABLE 7.3 (CONT.)

Saybolt viscosity, s^b				Kinematic viscosity, cST^b				Specific gravity 60/60°F (deg API), max.	Copper strip corro- sion, max.	Sulfur (%), max.
Universal at 38°C (100°F)		Furol at 50°C (122°F)		At 38°C (100°F)		At 50°C (122°F)				
min.	max.	min.	max.	min.	max.	min.	max.			
—	—	—	—	1.4	2.2	—	—	.8499 (35 min.)	No. 3	.5 or legal
(32.6)	(37.9)	—	—	2.0^c	3.6	—	—	.8762 (30 min.)	No. 3	$.5^g$ or legal
(45)	(125)	—	—	5.8	26.4^f	—	—	—	—	Legal
(>125)	(300)	—	—	>26.4	65^f	—	—	—	—	Legal
(>300)	(900)	(23)	(40)	>65	194^f	(42)	(81)	—	—	Legal
(>900)	(9000)	(>45)	(300)	>92	638^f	50°C	(122°F)	—	—	Legal

[f] Where low-sulfur fuel is required, fuel oil falling in the viscosity range of a lower-numbered grade down to and including No. 4 may be supplied by agreement between purchaser and supplier. The viscosity range of the initial shipment shall be identified and advance notice shall be required when changing from one viscosity range to another. This notice shall be in sufficient time to permit the user to make the necessary adjustments.

[g] In countries outside the United States other sulfur limits may apply.

Source: Reference 1. From *Combustion Fossil Power Systems,* Combustion Engineering, Inc. Used with permission. Excerpted from *ASTM Standards D 396,* Specification for Fuel Oils. Copyright, American Society for Testing and Materials, Philadelphia. Reprinted with permission.

7.2.2 Unconventional Fuels

Recently companies have begun to search for fuels less expensive than fuel oil, coal, and gas. A natural source for an inexpensive fuel is manufacturing waste, and this can be almost anything. Waste materials that have found use as fuels include pulp mill liquor, sawdust, food processing waste, municipal garbage, coal wash water, coffee grounds, cardboard, bark, and bagasse (sugar cane after the liquid has been extracted). Some sample analyses are given in Table 7–4, with natural gas included for comparison purposes. Using industrial waste as fuel can simplify the refuse disposal problem for a company as well as provide it with an inexpensive source of heat.

TABLE 7.4 COMPOSITION OF NONTRADITIONAL FUELS

Fuel	Sulfur (S)	Hydrogen (H$_2$)	Carbon (C)	Component oxygen (O$_2$)	Moisture (H$_2$O)	Ash	Heating value (Btu/lb)
Pine bark (dry basis)	.1%	5.6%	53.4%	37.9%	(50%)	2.9%	9,030
Natural gas	—	23.30	74.72	1.22	—	(.76% N$_2$)	22,904
Fuel oil No. 6	12.0	10.5	85.7	.92	2.0	.08	18,270
Coke breeze	.6	.3	80.0	.5	7.3	11.0	11,670
Bagasse	—	2.8	23.4	20.0	52.0	1.7	4,000
Municipal garbage (metals removed)	.1–.4	3.4–6.3	23.4–42.8	15.4–31.9	19.7–31.3	9.4–26.8	3100–6500

Source: References 1 and 2.

There are, however, some problems associated with burning any new fuel. Until the advent of fluidized-bed combustion, boilers were built around one or, at most, two fuels, and they usually required similar fuel transportation systems and burners. The technology for dealing with coal, gas, or fuel oil was well known. Using new fuels has, however, raised such questions as the following: (1) How high in the combustion chamber should the new fuel be injected into the boiler? (This is critical in burning municipal waste.) (2) What kind of problems will the ash or residue create? (3) What modifications are needed to burners and other boiler components? (4) What storage problems can be expected? (5) How regular will the supply be? In this technology, as with others, it is often best *not* to be the first one to try burning a new fuel. It also pays to have the best possible engineering advice.

One example of the costs to be calculated when considering a waste-fired boiler comes from Reference 3, a brief discussion of a Proctor and Gamble boiler and refuse-handling system installed to burn trash from a diaper and paper products factory. These costs and savings are listed in Table 7–5.

7.2.3 Cost Comparison Example

To illustrate some of the factors involved in a typical choice, consider the following example. The options are (1) to continue with an existing gas-fired boiler; (2) to construct a 200,000-lb/h boiler to burn municipal garbage and a 100,000-lb/h boiler for coal, using the coal-fired boiler only when the demand of the garbage-fired boiler is exceeded; or (3) to construct a 300,000-lb/h boiler fired by garbage alone, with a drying and storage facility to accommodate any surplus inventory of garbage that accumulates during some months of the year and must be held for use later in the year.

TABLE 7.5 WASTE-BURNING BOILER ECONOMICS, SAMPLE CASE

Savings (1980 dollars)	Costs (1980 dollars)

Coal and natural gas:
    $250,000/year
Trash hauling and landfill:
    $362,000/year
	Site preparation: $235,000
	Building to house system: $465,000
	Equipment support structures: $125,000
	Boiler and trash-handling equipment:
	    $1,250,000
	Piping: $200,000
	Instrumentation: $160,000
	Crew locker room: $140,000
	Miscellaneous mechanical equipment:
	    $85,000
	Spare parts: $40,000

Company management has decided to choose alternative 2 or 3 if its incremental constant-dollar after-tax return on investment is 15% or higher based on the first 10 years of cash flow. Fifteen percent is a low rate of return for this company and was chosen for this project because of its impact on living conditions in the local area.

Possible costs may include site preparation and construction, costs of modifying the gas boiler to accept coal, disposal of garbage to a landfill or preparation (drying) so that a reasonable supply can be kept on hand, ash haulage, transportation costs of the garbage from the town to the plant, waste storage and haulage, maintenance costs, and coal costs. All costs and dollar benefits except for those of gas are expected to increase at 8%/year; gas costs are expected to rise at 20%/year. Construction and site preparation costs for a new trash-fired boiler system will depend on its capacity (at 750 psi, the rating of the present system); for alternative 2, the garbage-fired boiler will cost $8,864,000 and the coal-fired boiler, $3,500,000; for alternative 3, the initial cost is $13 million. All boilers will qualify for the 10-year life depreciation tax schedule described in Chapter 5. Landfill costs are $5.00/ton of garbage or $2.00/ton of ash, including transportation costs to the landfill. Coal costs $55.00/ton and is expected to increase at 8% each year for the foreseeable future. Gas costs $5.00/Mcf (at 1 million Btu/Mcf, taking boiler efficiency into account) and is expected to increase at 20%/year. Repair and spare parts costs for alternative 2 are estimated as $350,000/year in addition to those presently incurred with gas. If the company chooses alternative 2 or 3, it is contractually obligated to buy all that the town presently produces, 219,000 tons/year, at $3.50/ton including transportation. If either alternative 2 or 3 is chosen, 17,000 tons/year of garbage must be hauled per year because of limited storage facilities. An analysis of the garbage shows it to contain 9% (by weight) recoverable metals and 16% ash and to have a heating value of 6390 Btu/lb as fired. The garbage-fired boiler is expected to operate at an efficiency of 75%. The coal to be used has a heating value of 12,780 Btu/lb and an ash content of 9.6%. The coal boiler efficiency is estimated as 82%, giving the effective heating value of coal as 21 million Btu/ton.

A major benefit from either alternative is the elimination of gas costs, presently $10,500,000/year. Other benefits include revenue from sales of scrap, with a long-term contract of $30/ton, plus inflation. This amount includes transportation. There is also a first-year tax credit of 15% of the initial cost. With these data, it is possible to estimate the annual costs for all three alternatives. The first-year costs for the gas alternative are $10,500,000. First-year costs for the coal and garbage combination can be calculated as shown in Figure 7–10.

Fuel costs = coal cost + garbage cost

$$\text{Coal cost} = \frac{12{,}000 \text{ tons garbage/year} \times .75 \times 12{,}780{,}000 \text{ btu/ton}}{21{,}000{,}000 \text{ btu/ton coal} \times .82} \times \$55/\text{ton coal}$$

$= \$520{,}440/\text{year}$

Garbage cost = 219,000 tons/year × $3.50/ton

$= \$766{,}500/\text{year}$ = $1,286,940/year

Ash haulage costs = coal ash + garbage ash

Coal ash cost = 9463 tons coal/year × .096 ton ash/ton coal ×
   $2.00/ton ash

$= \$1816/\text{year}$

Garbage ash cost = (219,000 − 17,00(0) tons garbage/year ×
   .16 ton ash/ton garbage × $2.00/ton ash

$= \$64{,}640/\text{year}$

Ash haulage cost   = $66,456/year

Wast haulage costs = 17,000 tons × $5.00/ton   = $85,000/year

Metal recovery benefits = (219,000 − 17,000) tons garbage/year ×
   .09 ton metal/ton garbage × $30.00/ton

$= \$545{,}400/\text{year}$

Metal recovery cost = −$545,400/year

Repair and spare parts = $350,000

Total: $1,243,000/year

Figure 7–10   Sample problem: annual cost calculations.

Alternative 3 requires a boiler capable of operating between 90,000 and 300,000 lb/h, well within the capacities of existing boilers. (The efficiency will generally be different at different steam rates, but the difference is neglected in this example.) For this new alternative, an estimate of the cost of building storage capacity is $2 million. The savings include the cost of the coal-fired boiler, costs of coal, and all costs associated with hauling raw garbage to a landfill. Some brief computations show these costs as follows: fuel, $766,500; ash haulage, $70,080; waste haulage, $.00; and metal recovery costs (benefits, hence negative), −$591,300. The total is $245,280.

With these data, it is possible to perform an economic analysis of the three alternatives. The relevant constant-dollar cash flows are shown in Figures 7–11(a), 7–11(b), and 7–11(c). An incremental analysis of these three cash flow streams is shown in Figure 7–12, where the net current-dollar cash flows have been compared for various values of the discount rate. An incremental analysis was used, as indicated, comparing first the coal and garbage boiler system to the gas system and then the garbage boiler alone to the coal and garbage system. Note that the break-even rate of return for the first comparison is about 82% and for the second comparison, about 33%. If the money is available, either the $12,140,000 investment or the $15 million investment would be worthwhile, and the high rate of return would be a potent argument in contention with nonquantitative factors.

Before leaving this chapter, it should be noted that other factors must be taken into account in the decision to use an alternate fuel. First is the need to have some kind of backup boiler if the waste fuel is not available. This problem is particularly

Year	Fuel costs	Tax deductions	Net current-dollar cash flow
0	—	—	—
1	$12,600	$5,796	$6,804
2	15,120	6,955	8,165
3	18,144	8,346	9,798
4	21,773	10,016	11,757
5	26,127	12,018	14,109
6	31,353	14,422	16,931
7	37,623	17,307	20,316
8	45,148	20,768	24,380
9	54,178	24,922	29,256
10	65,013	29,906	35,107

(a)

Figure 7–11(a) Sample problem: current-dollar costs with present system (costs in $ × 1000).

Year	Capital	Fuel et al.	Depreciation	Taxes	Net cash flow
0	$12,140	—	—	−$1821	$10,319
1	—	$1342	($971)	−1064	278
2		1450	(1700)	−1449	1
3		1566	(1457)	−1391	175
4		1691	(1214)	−1336	355
5		1826	(1214)	−1398	428
6		1972	(1214)	−1466	506
7		2130	(1093)	−1483	647
8		2301	(1093)	−1561	740
9		2405	(1093)	−1646	839
10		2684	(1093)	−1737	947

(b)

Figure 7–11(b) Sample problem: current-dollar costs with garbage and coal boilers ($ × 1000).

Year	Capital	Fuel et al.	Depreciation	Taxes	Net cash flow
0	$15,000	—	—	$−2250	$12,750
1	—	$265	($1200)	−674	−409
2		286	(2100)	−1098	−812
3		194	(1800)	−917	−723
4		170	(1500)	−768	−598
5		146	(1500)	−757	−611
6		121	(1500)	−746	−525
7		97	(1350)	−666	−569
8		73	(1350)	−655	−582
9		49	(1350)	−644	−595
10		24	(1350)	−632	−608

(c)

Figure 7–11(c)   Sample problem: current-dollar costs with garbage boiler alone ($ × 1000).

Discount rate	Present worth of coal and garbage minus gas	Present worth of garbage alone minus coal and garbage
0%	$127,229	$696
2	109,394	654
4	94,440	612
6	81,840	569
8	71,175	526
10	62,108	483
15	44,764	377
20	32,792	273
30	18,234	76
40	10,394	−205
50	5,881	−261
60	3,137	−402
70	1,395	−526
80	250	−637
90	−1,047	−736
100	−1,056	−824

Figure 7–12 Sample problem: comparison of the three alternatives (current $ × 1000).

severe when one company uses the waste that is produced by another company; it was also responsible for some of the early difficulties experienced in the attempt to use municipal refuse as a fuel. A second major factor is the political climate. It is necessary to determine what governmental agencies must give their approval before a particular plan can be put into effect. In the case of municipal refuse, political problems probably delayed more projects than did technological difficulties. The problem of refuse storage is also critical; see Problem 7.6 for an example.

7.3 SUMMARY

In this chapter we have presented the reactions and the standard analysis techniques needed for a fundamental understanding of combustion. Specific techniques discussed were weight balances, heat balances, and flue gas analysis. These techniques were used to demonstrate the importance of proper control of air quantities for combustion of common fuels and as a background for the discussion of unconventional fuels. An example was presented that showed some of the analysis steps and problems to be considered when considering municipal garbage as a fuel source.

REFERENCES

1. Joseph G. Singer, *Combustion/Fossil Power Systems,* Combustion Engineering, Inc., Windsor, Conn., 1981.
2. *Steam/Its Generation and Use,* Babcock and Wilcox, New York, 1978.
3. Cindy Galvin, ''Trash-Fired Boiler Nets Paper Firm $600,000 Savings,'' *Energy User News,* Vol. 7, No. 4, p. 8, Jan. 25, 1982.

BIBLIOGRAPHY

SCHROEDER, GEORGE, *Combustion and Energy Loss in Boilers,* Mountain Fuel Supply Company, Salt Lake City, 1977.

SCOLLON, R. B., and R. D. SMITH, ''Boilers and Fired Systems,'' in *Energy Management Handbook,* Wayne C. Turner, Senior ed., Wiley-Interscience, New York, 1982, Chap. 5.

QUESTIONS

7.1. What chemical processes make up a flame? Take one gas, say CH^4, and show as many steps as you can. Include the formation of free radicals in your explanation.

7.2. What are NO_x, and SO_x, and how are they formed in a boiler?

7.3. What concentration of CO is poisonous to humans? How can this answer change the desirability of excess air?

7.4. In Section 7.2.2, a series of questions were asked to illustrate the uncertainties associated with burning any new fuel. Make a list of 10 such questions on the economics of burning the industrial waste most prevalent in your area.

7.5. Many of the basic costs given in the example of Section 7.2.3 are subject to change. How do you (a) estimate the range of parameter values, (b) express the resulting economic evaluation in ways that are clear to the public, and (c) incorporate the range of values into your decision-making process?

7.6. What could go wrong with the garbage-fired boiler proposal, and what additional data do you need now to prevent these problems?

7.7. What is a good value for the minimum attractive rate of return for use in choosing a waste-fired boiler?

PROBLEMS

7.1. A refinery gas available as a fuel has the following dry chemical composition, by volume [1]:

CO_2:	3.3%	C_2H_6:	19.8%
CO:	1.5%	C_3H_8:	38.1%
H_2:	5.6%	C_4H_{10}:	.8%
CH_4:	30.9%		

(a) Determine the heating value in Btu/ft^3 and in Btu/lb.

(b) Assuming that 15% excess air is needed for complete combustion, determine the total amount of combustion air needed for each cubic foot of this gas.

7.2. For the gas in Problem 7.1, assuming that 15% excess air is needed for complete combustion, determine the flue gas composition (a) by volume and (b) by weight.

7.3. A particular Utah coal has the following proximate and ultimate analyses:

Proximate analysis		Ultimate analysis	
Component	Weight (%)	Component	Weight (%)
Moisture	4.3	Moisture	4.3
Volatile matter	37.2	Carbon	72.2
Fixed carbon	51.8	Hydrogen	5.1
Ash	6.7	Sulfur	1.1
	Total: 100.0	Nitrogen	1.6
		Oxygen	9.0
Heating value: 12,990 Btu/lb		Ash	6.7
			Total: 100.0

(a) Determine the amount of air needed for combustion, assuming complete mixing.

(b) Calculate the flue gas composition by weight, assuming that 25% excess combustion air is needed and that standard air (relative humidity, 60%; temperature, 80°F) is used.

(c) Assuming that the boiler produces 3 million Btu/h for 4000 h each year and that the flue gas temperature is 750°F, determine the amount and cost of coal using $60.00/ton as the delivered coal cost.

7.4. Suppose that a respectable firm has been advertising a new burner system that would enable the boiler in Section 7.1.2 to operate at 8% excess air before CO was detected in the flue gas. How much would this increase the boiler efficiency?

7.5. In Section 7.2.2, a waste-burning boiler was described. The capacity of this boiler was (from Reference 3) 28,000 lb/h. Suppose that these figures are 5 years old, that your company is contemplating the purchase of such a boiler, and that it is planned to save twice the energy amounts and have twice the capacity of the given boiler. The energy cost has been inflating at 20%/year, base construction costs have been inflating at 6%/year, the basic inflation rate of the economy has been 5%, and without inflation the cost of constructing a unit is $R^{.73}$ multiplied by the cost of the existing unit, where R is the ratio between the capacity of the proposed unit and the capacity of the present unit. The tax rate of the company is 48%. The unit creates a 10% tax credit the first year of operation and is subject to a 5-year depreciation schedule. What is the after-tax present worth of the first 5 years of cash flows associated with this investment if the company uses a constant-dollar after-tax rate of return of 8% on this kind of investment?

7.6. The choice of an optimum combination of boiler sizes in the garbage-coal situation is not usually easy. Suppose that health conditions limit the time that garbage, even dried, can be stored to 1 month, that the initial costs given in the accompanying table are valid, and that the municipality and your company have supplies and needs for energy, respectively, as given in the table labeled "data" for Problem 7.6. Suppose that all the other costs for this problem are the same as in Section 7.2.3. What is the optimum choice now?

Costs for Problem 7.6

Capacity, 750 psi	Initial costs: trash-fired boiler	Initial costs: coal-fired boiler
50,000 lb/h	—	$1,800,000
100,000	—	3,500,000
150,000	$6,250,000	5,100,000
200,000	8,640,000	6,900,000
250,000	10,870,000	8,900,000
300,000	13,000,000	11,000,000

Data for Problem 7.6

Month	Garbage needed (tons)	Garbage available (tons)
January	23,000	13,500
February	23,000	13,500
March	21,600	16,500
April	19,500	18,000
May	14,100	18,900
June	9,500	19,500
July	7,600	22,500
August	9,500	21,000
September	10,800	21,000
October	13,500	18,000
November	18,400	15,000
December	24,300	18,600

Chapter 8

Steam Generation
and Distribution

8.0 INTRODUCTION

Most fuel purchased is used to generate steam. Steam is then used to generate electricity or distributed to provide comfort or process heat. Many significant opportunities for energy cost reduction can be found in a good technical examination of these steam generation and distribution systems. Such an examination first gives an estimate of the amount of energy coming into a facility and how it leaves. These estimates then provide a guide to possible waste heat utilization and can be used to evaluate insulation possibilities. At the same time, maintenance is considered in detail. If it seems appropriate, cogeneration is also examined for feasibility. All these factors are important and are described in this text. Maintenance and insulation of these systems are covered in Chapters 10 and 11, respectively. The present chapter provides an analytical framework for those chapters and discusses in greater detail boilers, waste heat recovery from boilers and steam distribution systems, and cogeneration.

8.1 ANALYZING ENERGY CONSUMPTION

One of the most pervasive principles of analysis is that the most effort should be placed where the most opportunity lies. In relating this principle to boiler operation, analyzing for opportunity consists of two parts. First, a heat balance is performed to show where the boiler energy comes from—the energy sources—and where it goes—the energy sinks. The energy sources are ranked in order of the amount of energy they supply, and the energy sinks are ranked by the amount of heat they exhaust to the environment. Then the source components are examined to see whether each of their functions can be performed with less energy, and the sink components are scrutinized to see if they can give up their energy in some useful way to some source component. The result of this process is a knowledge of where energy is utilized or wasted in the boiler and steam distribution system. This knowledge then provides a measure of the benefit to be gained from insulation and from waste heat utilization.

8.1.1 The Heat Balance

The fundamental tool used in the analysis of boilers and steam distribution systems is the heat balance. A heat balance attempts to account for all the heat that goes into a system and to find where all of it leaves the system. This kind of principle was used in the mass balance of chemical reactions in Chapter 7, and it is very useful. Before it can be used, however, it is necessary to obtain a better understanding of the heat content of steam, a concept embodied in the term *enthalpy*. Then basic principles of heat transfer are explained and quantified. These are followed by several examples of heat balances.

8.1.1.1 Enthalpy.   It takes about 970 Btu to convert 1 lb of water to steam at atmospheric pressure and at 212°F. This energy, called *latent* heat, is given up whenever the steam condenses. Steam also carries *sensible* heat, heat proportional to the temperature difference through which the steam was heated. Steam under pressure contains additional heat due to the mechanical work done on the vapor. (The latent heat is generally lower at higher pressures.) The sum of the latent heat, the sensible heat, and the mechanical work is called *enthalpy* and, when expressed in Btu/lb, *specific enthalpy*. (This definition of enthalpy does not take into account the internal kinetic energy of the steam, but this can usually be neglected in energy management [1].) Where water is present with the steam, as in most steam distribution systems, the steam is said to be *saturated,* and the pressure increases as the temperature increases. Furthermore the specific volume, ft³/lb, decreases as the temperature increases. The enthalpy and specific volume for various temperature and pressure conditions are shown in Table 8–1.

Table 8–1 has many uses. For example, suppose that a steam leak is estimated to be losing 150 lb of steam per hour from a 1200-psi line, to an ambient temperature of 70°F. From Table 8–1, the enthalpy at 1200 psi is 1183.4 Btu/lb and at 70°F, 70 − 32, or 38 Btu/lb. (From 70 to 212°F, the water is all liquid, and the only heat is sensible. Raising the temperature of 1 lb of *water* 1°F takes 1 Btu.) Thus the total change in enthalpy is 1183 − 38, or 1145 Btu/lb. A loss of 150 lb/h represents a loss of 171,750 Btu/h. If fuel costs $8.00/million Btu, this represents a cost of $1.37/h or $33/day on days when the facility is running constantly. The amount of *flash steam* (steam released to the atmosphere) released by this leak is 26.80 × 150, or 4020 ft³/h.

TABLE 8.1 PROPERTIES OF SATURATED STEAM AND SATURATED WATER

Temp F	Press. psia	Volume, ft³/lb			Enthalpy, Btu/lb			Entropy, Btu/lb, F			Temp F
		Water v_f	Evap v_{fg}	Steam v_g	Water h_f	Evap h_{fg}	Steam h_g	Water s_f	Evap s_{fg}	Steam s_g	
32	0.08859	0.01602	3305	3305	−0.02	1075.5	1075.5	0.0000	2.1873	2.1873	32
35	0.09991	0.01602	2948	2948	3.00	1073.8	1076.8	0.0061	2.1706	2.1767	35
40	0.12163	0.01602	2446	2446	8.03	1071.0	1079.0	0.0162	2.1432	2.1594	40
45	0.14744	0.01602	2037.7	2037.8	13.04	1068.1	1081.2	0.0262	2.1164	2.1426	45
50	0.17796	0.01602	1704.8	1704.8	18.05	1065.3	1083.4	0.0361	2.0901	2.1262	50
60	0.2561	0.01603	1207.6	1207.6	28.06	1059.7	1087.7	0.0555	2.0391	2.0946	60
70	0.3629	0.01605	868.3	868.4	38.05	1054.0	1092.1	0.0745	1.9900	2.0645	70
80	0.5068	0.01607	633.3	633.3	48.04	1048.4	1096.4	0.0932	1.9426	2.0359	80
90	0.6981	0.01610	468.1	468.1	58.02	1042.7	1100.8	0.1115	1.8970	2.0086	90
100	0.9492	0.01613	350.4	350.4	68.00	1037.1	1105.1	0.1295	1.8530	1.9825	100
110	1.2750	0.01617	265.4	265.4	77.98	1031.4	1109.3	0.1472	1.8105	1.9577	110
120	1.6927	0.01620	203.25	203.26	87.97	1025.6	1113.6	0.1646	1.7693	1.9339	120
130	2.2230	0.01625	157.32	157.33	97.96	1019.8	1117.8	0.1817	1.7295	1.9112	130
140	2.8892	0.01629	122.98	123.00	107.95	1014.0	1122.0	0.1985	1.6910	1.8895	140
150	3.718	0.01634	97.05	97.07	117.95	1008.2	1126.1	0.2150	1.6536	1.8686	150
160	4.741	0.01640	77.27	77.29	127.96	1002.2	1130.2	0.2313	1.6174	1.8487	160
170	5.993	0.01645	62.04	62.06	137.97	996.2	1134.2	0.2473	1.5822	1.8295	170
180	7.511	0.01651	50.21	50.22	148.00	990.2	1138.2	0.2631	1.5480	1.8111	180
190	9.340	0.01657	40.94	40.96	158.04	984.1	1142.1	0.2787	1.5148	1.7934	190
200	11.526	0.01664	33.62	33.64	168.00	977.9	1146.0	0.2940	1.4824	1.7764	200
210	14.123	0.01671	27.80	27.82	178.15	971.6	1149.7	0.3091	1.4509	1.7600	210
212	14.696	0.01672	26.78	26.80	180.17	970.3	1150.5	0.3121	1.4447	1.7568	212
220	17.186	0.01678	23.13	23.15	188.23	965.2	1153.4	0.3241	1.4201	1.7442	220
230	20.779	0.01685	19.364	19.381	198.33	958.7	1157.1	0.3388	1.3902	1.7290	230
240	24.968	0.01693	16.304	16.321	208.45	952.1	1160.6	0.3533	1.3609	1.7142	240
250	29.825	0.01701	13.802	13.819	218.59	945.4	1164.0	0.3677	1.3323	1.7000	250
260	35.427	0.01709	11.745	11.762	228.76	938.6	1167.4	0.3819	1.3043	1.6862	260
270	41.856	0.01718	10.042	10.060	238.95	931.7	1170.6	0.3960	1.2769	1.6729	270
280	49.200	0.01726	8.627	8.644	249.17	924.6	1173.8	0.4098	1.2501	1.6599	280
290	57.550	0.01736	7.443	7.460	259.4	917.4	1176.8	0.4236	1.2238	1.6473	290
300	67.005	0.01745	6.448	6.466	269.7	910.0	1179.7	0.4372	1.1979	1.6351	300
310	77.67	0.01755	5.609	5.626	280.0	902.5	1182.5	0.4506	1.1726	1.6232	310
320	89.64	0.01766	4.896	4.914	290.4	894.8	1185.2	0.4640	1.1477	1.6116	320
340	117.99	0.01787	3.770	3.788	311.3	878.8	1190.1	0.4902	1.0990	1.5892	340
360	153.01	0.01811	2.939	2.957	332.3	862.1	1194.4	0.5161	1.0517	1.5678	360
380	195.73	0.01836	2.317	2.335	353.6	844.5	1198.0	0.5416	1.0057	1.5473	380
400	247.26	0.01864	1.8444	1.8630	375.1	825.9	1201.0	0.5667	0.9607	1.5274	400
420	308.78	0.01894	1.4808	1.4997	396.9	806.2	1203.1	0.5915	0.9165	1.5080	420
440	381.54	0.01926	1.1976	1.2169	419.0	785.4	1204.4	0.6161	0.8729	1.4890	440
460	466.9	0.0196	0.9746	0.9942	441.5	763.2	1204.8	0.6405	0.8299	1.4704	460
480	566.2	0.0200	0.7972	0.8172	464.5	739.6	1204.1	0.6648	0.7871	1.4518	480
500	680.9	0.0204	0.6545	0.6749	487.9	714.3	1202.2	0.6890	0.7443	1.4333	500
520	812.5	0.0209	0.5386	0.5596	512.0	687.0	1199.0	0.7133	0.7013	1.4146	520
540	962.8	0.0215	0.4437	0.4651	536.8	657.5	1194.3	0.7378	0.6577	1.3954	540
560	1133.4	0.0221	0.3651	0.3871	562.4	625.3	1187.7	0.7625	0.6132	1.3757	560
580	1326.2	0.0228	0.2994	0.3222	589.1	589.9	1179.0	0.7876	0.5673	1.3550	580
600	1543.2	0.0236	0.2438	0.2675	617.1	550.6	1167.7	0.8134	0.5196	1.3330	600
620	1786.9	0.0247	0.1962	0.2208	646.9	506.3	1153.2	0.8403	0.4689	1.3092	620
640	2059.9	0.0260	0.1543	0.1802	679.1	454.6	1133.7	0.8686	0.4134	1.2821	640
660	2365.7	0.0277	0.1166	0.1443	714.9	392.1	1107.0	0.8995	0.3502	1.2498	660
680	2708.6	0.0304	0.0808	0.1112	758.5	310.1	1068.5	0.9365	0.2720	1.2086	680
700	3094.3	0.0366	0.0386	0.0752	822.4	172.7	995.2	0.9901	0.1490	1.1390	700
705.5	3208.2	0.0508	0	0.0508	906.0	0	906.0	1.0612	0	1.0612	705.5

Source: Abstracted from *Thermodynamic and Transport Properties of Steam,* © 1967 by the American Society of Mechanical Engineers.

In some applications, steam is present without water. Such steam is said to be *superheated,* and its enthalpy and specific volume at various temperatures and pressures are give in Table 8–2.

Example 1

If a throttling valve reduces the temperature and pressure of superheated steam from 1300°F and 1200 psi to 700°F and 600 psi, how much work is being lost?

Solution. The difference in enthalpy is 1671.6 − 1351.8, or 319.8 Btu/lb. This represents heat that could possibly be used in some other part of the facility.

8.1.1.2 Heat losses. Enthalpy is a useful way of describing the amount of energy contained in steam or in steam condensate. Much of the energy lost from boilers, however, is radiated to the environment, and it is important to know how to calculate this loss so as to complete the heat balance. A knowledge of heat transfer is also useful in determining the value of steam traps.

First, consider the problem of heat losses from hot pipes or boiler surfaces. These heat losses are associated with convection (usually treated as if it were the free convection associated with still air) and with radiation. This kind of heat loss must be considered whenever boilers or piping are contained within the system being analyzed, and it must be calculated for all surfaces. Useful formulas for these heat losses are given in Reference 2:

$$\text{radiative loss} = A \times .1714 \times 10^{-8} \times (T_S^4 - T_R^4) \tag{8-1}$$

and

$$\text{convective loss} = A \times .18 \times (T_S - T_R)^{4/3} \tag{8-2}$$

where A = surface area in ft^2
$.1714 \times 10^{-8}$ = Stefan-Boltzmann constant in Btu/h·ft^2·°R^4
T_S, T_R = surface and room temperatures, respectively, in °R (°F + 460)

Example 2

Calculate the amount of heat loss from the 140°F surface of a boiler into a room whose temperature is 85°F. The exposed surface of the boiler is 400 ft^2.

Solution. The radiative loss, per square foot per hour, is given by Equation 8–1 as $1 \times .1714 \times 10^{-8} \times [(140 + 460)^4 - (80 + 460)^4]$ or 76.4 Btu/ft^2. The convective loss, per square foot, is given by Equation 8–2 as $.18 \times 60^{4/3}$, or 42.2 Btu/ft^2. Thus total heat loss due to convection and radiation is $400 \times (42.2 + 76.4)$, or 47,440 Btu/h.

A second major heat loss occurs in the energy removed in the steam itself. The enthalpy is given in Tables 8–1 and 8–2 for various steam temperatures; the flow rate and temperature can be observed as readings on gauges mounted on the steam lines. For example, the amount of energy flowing through a pipe with steam at 250 psig at a rate of 600 lb/h is 600 lb/h × 1201.7 Btu/lb, or 720,000 Btu/h.

Any other material leaving the system must be taken into account in the energy balance. For example, flue gas carries with it both latent and sensible heat, as was explained in Chapter 7, and the energy lost this way must be calculated as a significant

TABLE 8.2 PROPERTIES OF SUPERHEATED STEAM AND COMPRESSED WATER

Abs press. lb/sq in. (sat. temp)		100	200	300	400	500	600	700	800	900	1000	1100	1200	1300	1400	1500
1 (101.74)	v	0.0161	392.5	452.3	511.9	571.5	631.1	690.7								
	h	68.00	1150.2	1195.7	1241.8	1288.6	1336.1	1384.5								
	s	0.1295	2.0509	2.1152	2.1722	2.2237	2.2708	2.3144								
5 (162.24)	v	0.0161	78.14	90.24	102.24	114.21	126.15	138.08	150.01	161.94	173.86	185.78	197.70	209.62	221.53	233.45
	h	68.01	1148.6	1194.8	1241.3	1288.2	1335.9	1384.3	1433.6	1483.7	1534.7	1586.7	1639.6	1693.3	1748.0	1803.5
	s	0.1295	1.8716	1.9369	1.9943	2.0460	2.0932	2.1369	2.1776	2.2159	2.2521	2.2866	2.3194	2.3509	2.3811	2.4101
10 (193.21)	v	0.0161	38.84	44.98	51.03	57.04	63.03	69.00	74.98	80.94	86.91	92.87	98.84	104.80	110.76	116.72
	h	68.02	1146.6	1193.7	1240.6	1287.8	1335.5	1384.0	1433.4	1483.5	1534.6	1586.6	1639.5	1693.3	1747.9	1803.4
	s	0.1295	1.7928	1.8593	1.9173	1.9692	2.0166	2.0603	2.1011	2.1394	2.1757	2.2101	2.2430	2.2744	2.3046	2.3337
15 (213.03)	v	0.0161	0.0166	29.899	33.963	37.985	41.986	45.978	49.964	53.946	57.926	61.905	65.882	69.858	73.833	77.807
	h	68.04	168.09	1192.5	1239.9	1287.3	1335.2	1383.8	1433.2	1483.4	1534.5	1586.5	1639.4	1693.2	1747.8	1803.4
	s	0.1295	0.2940	1.8134	1.8720	1.9242	1.9717	2.0155	2.0563	2.0946	2.1309	2.1653	2.1982	2.2297	2.2599	2.2890
20 (227.96)	v	0.0161	0.0166	22.356	25.428	28.457	31.466	34.465	37.458	40.447	43.435	46.420	49.405	52.388	55.370	58.352
	h	68.05	168.11	1191.4	1239.2	1286.9	1334.9	1383.5	1432.9	1483.2	1534.3	1586.3	1639.3	1693.1	1747.8	1803.3
	s	0.1295	0.2940	1.7805	1.8397	1.8921	1.9397	1.9836	2.0244	2.0628	2.0991	2.1336	2.1665	2.1979	2.2282	2.2572
40 (267.25)	v	0.0161	0.0166	11.036	12.624	14.165	15.685	17.195	18.699	20.199	21.697	23.194	24.689	26.183	27.676	29.168
	h	68.10	168.15	1186.6	1236.4	1285.0	1333.6	1382.5	1432.1	1482.5	1533.7	1585.8	1638.8	1992.7	1747.5	1803.0
	s	0.1295	0.2940	1.6992	1.7608	1.8143	1.8624	1.9065	1.9476	1.9860	2.0224	2.0569	2.0899	2.1224	2.1516	2.1807
60 (292.71)	v	0.0161	0.0166	7.257	8.354	9.400	10.425	11.438	12.446	13.450	14.452	15.452	16.450	17.448	18.445	19.441
	h	68.15	168.20	1181.6	1233.5	1283.2	1332.3	1381.5	1431.3	1481.8	1533.2	1585.3	1638.4	1692.4	1747.1	1802.8
	s	0.1295	0.2939	1.6492	1.7134	1.7681	1.8168	1.8612	1.9024	1.9410	1.9774	2.0120	2.0450	2.0765	2.1068	2.1359
80 (312.04)	v	0.0161	0.0166	0.0175	6.218	7.018	7.794	8.560	9.319	10.075	10.829	11.581	12.331	13.081	13.829	14.577
	h	68.21	168.24	269.74	1230.5	1281.3	1330.9	1380.5	1430.5	1481.1	1532.6	1584.9	1638.0	1692.0	1746.8	1802.5
	s	0.1295	0.2939	0.4371	1.6790	1.7349	1.7842	1.8289	1.8702	1.9089	1.9454	1.9800	2.0131	2.0446	2.0750	2.1041
100 (327.82)	v	0.0161	0.0166	0.0175	4.935	5.588	6.216	6.833	7.443	8.050	8.655	9.258	9.860	10.460	11.060	11.659
	h	68.26	168.29	269.77	1227.4	1279.3	1329.6	1379.5	1429.7	1480.4	1532.0	1584.4	1637.6	1691.6	1746.5	1802.2
	s	0.1295	0.2939	0.4371	1.6516	1.7088	1.7586	1.8036	1.8451	1.8839	1.9205	1.9552	1.9883	2.0199	2.0502	2.0794
120 (341.27)	v	0.0161	0.0166	0.0175	4.0786	4.6341	5.1637	5.6831	6.1928	6.7006	7.2060	7.7096	8.2119	8.7130	9.2134	9.7130
	h	68.31	168.33	269.81	1224.1	1277.4	1328.1	1378.4	1428.8	1479.8	1531.4	1583.9	1637.1	1691.3	1746.2	1802.0
	s	0.1295	0.2939	0.4371	1.6286	1.6872	1.7376	1.7829	1.8246	1.8635	1.9001	1.9349	1.9680	1.9996	2.0300	2.0592
140 (353.04)	v	0.0161	0.0166	0.0175	3.4661	3.9526	4.4119	4.8585	5.2995	5.7364	6.1709	6.6036	7.0349	7.4652	7.8946	8.3233
	h	68.37	168.38	269.85	1220.8	1275.3	1326.8	1377.4	1428.0	1479.1	1530.8	1583.4	1636.7	1690.9	1745.9	1801.7
	s	0.1295	0.2939	0.4370	1.6085	1.6686	1.7196	1.7652	1.8071	1.8461	1.8828	1.9176	1.9508	1.9825	2.0129	2.0421
160 (363.55)	v	0.0161	0.0166	0.0175	3.0060	3.4413	3.8480	4.2420	4.6295	5.0132	5.3945	5.7741	6.1522	6.5293	6.9055	7.2811
	h	68.42	168.42	269.89	1217.4	1273.3	1325.4	1376.4	1427.2	1478.4	1530.3	1582.9	1636.3	1690.5	1745.6	1801.4
	s	0.1294	0.2938	0.4370	1.5906	1.6522	1.7039	1.7499	1.7919	1.8310	1.8678	1.9027	1.9359	1.9676	1.9980	2.0273
180 (373.08)	v	0.0161	0.0166	0.0174	2.6474	3.0433	3.4093	3.7621	4.1084	4.4505	4.7907	5.1289	5.4657	5.8014	6.1363	6.4704
	h	68.47	168.47	269.92	1213.8	1271.2	1324.0	1375.3	1426.3	1477.7	1529.7	1582.4	1635.9	1690.2	1745.3	1801.2
	s	0.1294	0.2938	0.4370	1.5743	1.6376	1.6900	1.7362	1.7784	1.8176	1.8545	1.8894	1.9227	1.9545	1.9849	2.0142
200 (381.80)	v	0.0161	0.0166	0.0174	2.3598	2.7247	3.0583	3.3783	3.6915	4.0008	4.3077	4.6128	4.9165	5.2191	5.5209	5.8219
	h	68.52	168.51	269.96	1210.1	1269.0	1322.6	1374.3	1425.5	1477.0	1529.1	1581.9	1635.4	1689.8	1745.0	1800.9
	s	0.1294	0.2938	0.4369	1.5593	1.6242	1.6776	1.7239	1.7663	1.8057	1.8426	1.8776	1.9109	1.9427	1.9732	2.0025
250 (400.97)	v	0.0161	0.0166	0.0174	0.0186	2.1504	2.4662	2.6872	2.9410	3.1909	3.4382	3.6837	3.9278	4.1709	4.4131	4.6546
	h	68.66	168.63	270.05	375.10	1263.5	1319.0	1371.6	1423.4	1475.3	1527.6	1580.6	1634.4	1688.9	1744.2	1800.2
	s	0.1294	0.2937	0.4368	0.5667	1.5951	1.6502	1.6976	1.7405	1.7801	1.8173	1.8524	1.8858	1.9177	1.9482	1.9776
300 (417.35)	v	0.0161	0.0166	0.0174	0.0186	1.7665	2.0044	2.2263	2.4407	2.6509	2.8585	3.0643	3.2688	3.4721	3.6746	3.8764
	h	68.79	168.74	270.14	375.15	1257.7	1315.2	1368.9	1421.3	1473.6	1526.2	1579.4	1633.3	1688.0	1743.4	1799.6
	s	0.1294	0.2937	0.4307	0.5665	1.5703	1.6274	1.6758	1.7192	1.7591	1.7964	1.8317	1.8652	1.8972	1.9278	1.9572
350 (431.73)	v	0.0161	0.0166	0.0174	0.0186	1.4913	1.7028	1.8970	2.0832	2.2652	2.4445	2.6219	2.7980	2.9730	3.1471	3.3205
	h	68.92	168.85	270.24	375.21	1251.5	1311.4	1366.2	1419.2	1471.8	1524.7	1578.2	1632.3	1687.1	1742.6	1798.9
	s	0.1293	0.2936	0.4367	0.5664	1.5483	1.6077	1.6571	1.7009	1.7411	1.7787	1.8141	1.8477	1.8798	1.9105	1.9400
400 (444.60)	v	0.0161	0.0166	0.0174	0.0162	1.2841	1.4763	1.6499	1.8151	1.9759	2.1339	2.2901	2.4450	2.5987	2.7515	2.9037
	h	69.05	168.97	270.33	375.27	1245.1	1307.4	1363.4	1417.0	1470.1	1523.3	1576.9	1631.2	1686.2	1741.9	1798.2
	s	0.1293	0.2935	0.4366	0.5663	1.5282	1.5901	1.6406	1.6850	1.7255	1.7632	1.7988	1.8325	1.8647	1.8955	1.9250
500 (467.01)	v	0.0161	0.0166	0.0174	0.0186	0.9919	1.1584	1.3037	1.4397	1.5708	1.6992	1.8256	1.9507	2.0746	2.1977	2.3200
	h	69.32	169.19	270.51	375.38	1231.2	1299.1	1357.7	1412.7	1466.6	1520.3	1574.4	1629.1	1684.4	1740.3	1796.9
	s	0.1292	0.2934	0.4364	0.5660	1.4921	1.5595	1.6123	1.6578	1.6990	1.7371	1.7730	1.8069	1.8393	1.8702	1.8998

TABLE 8.2 (*CONT.*)

Abs press. lb/sq in. (sat. temp)		100	200	300	400	500	600	700	800	900	1000	1100	1200	1300	1400	1500
									Temperature, F							
600 (486.20)	v	0.0161	0.0166	0.0174	0.0186	0.7944	0.9456	1.0726	1.1892	1.3008	1.4093	1.5160	1.6211	1.7252	1.8284	1.9309
	h	69.58	169.42	270.70	375.49	1215.9	1290.3	1351.8	1408.3	1463.0	1517.4	1571.9	1627.0	1682.6	1738.8	1795.6
	s	0.1292	0.2933	0.4362	0.5657	1.4590	1.5329	1.5844	1.6351	1.6769	1.7155	1.7517	1.7859	1.8184	1.8494	1.8792
700 (503.08)	v	0.0161	0.0166	0.0174	0.0186	0.0204	0.7928	0.9072	1.0102	1.1078	1.2023	1.2948	1.3858	1.4757	1.5647	1.6530
	h	69.84	169.65	270.89	375.61	487.93	1281.0	1345.6	1403.7	1459.4	1514.4	1569.4	1624.8	1680.7	1737.2	1794.3
	s	0.1291	0.2932	0.4360	0.5655	0.6889	1.5090	1.5673	1.6154	1.6580	1.6970	1.7335	1.7679	1.8006	1.8318	1.8617
800 (518.21)	v	0.0161	0.0166	0.0174	0.0186	0.0204	0.6774	0.7828	0.8759	0.9631	1.0470	1.1289	1.2093	1.2885	1.3669	1.4446
	h	70.11	169.88	271.07	375.73	487.88	1271.1	1339.2	1399.1	1455.8	1511.4	1566.9	1622.7	1678.9	1735.0	1792.9
	s	0.1290	0.2930	0.4358	0.5652	0.6885	1.4869	1.5484	1.5980	1.6413	1.6807	1.7175	1.7522	1.7851	1.8164	1.8464
900 (531.95)	v	0.0161	0.0166	0.0174	0.0186	0.0204	0.5869	0.6858	0.7713	0.8504	0.9262	0.9998	1.0720	1.1430	1.2131	1.2825
	h	70.37	170.10	271.26	375.84	487.83	1260.6	1332.7	1394.4	1452.2	1508.5	1564.4	1620.6	1677.1	1734.1	1791.6
	s	0.1290	0.2929	0.4357	0.5649	0.6881	1.4659	1.5311	1.5822	1.6263	1.6662	1.7033	1.7382	1.7713	1.8028	1.8329
1000 (544.58)	v	0.0161	0.0166	0.0174	0.0186	0.0204	0.5137	0.6080	0.6875	0.7603	0.8295	0.8966	0.9622	1.0266	1.0901	1.1529
	h	70.63	170.33	271.44	375.96	487.79	1249.3	1325.9	1389.6	1448.5	1504.4	1561.9	1618.4	1675.3	1732.5	1790.3
	s	0.1289	0.2928	0.4355	0.5647	0.6876	1.4457	1.5149	1.5677	1.6126	1.6530	1.6905	1.7256	1.7589	1.7905	1.8207
1100 (556.28)	v	0.0161	0.0166	0.0174	0.0185	0.0203	0.4531	0.5440	0.6188	0.6865	0.7505	0.8121	0.8723	0.9313	0.9894	1.0468
	h	70.90	170.56	271.63	376.08	487.75	1237.3	1318.8	1384.7	1444.7	1502.4	1559.4	1616.3	1673.5	1731.0	1789.0
	s	0.1289	0.2927	0.4353	0.5644	0.6872	1.4259	1.4996	1.5542	1.6000	1.6410	1.6787	1.7141	1.7475	1.7793	1.8097
1200 (567.19)	v	0.0161	0.0166	0.0174	0.0185	0.0203	0.4016	0.4905	0.5615	0.6250	0.6845	0.7418	0.7974	0.8519	0.9055	0.9584
	h	71.16	170.78	271.82	376.20	487.72	1224.2	1311.5	1379.7	1440.9	1499.4	1556.9	1614.2	1671.6	1729.4	1787.6
	s	0.1288	0.2926	0.4351	0.5642	0.6868	1.4061	1.4851	1.5415	1.5883	1.6298	1.6679	1.7035	1.7371	1.7691	1.7996
1400 (587.07)	v	0.0161	0.0166	0.0174	0.0185	0.0203	0.3176	0.4059	0.4712	0.5282	0.5809	0.6311	0.6798	0.7272	0.7737	0.8195
	h	71.68	171.24	272.19	376.44	487.65	1194.1	1296.1	1369.3	1433.2	1493.2	1551.8	1609.9	1668.0	1727.0	1785.0
	s	0.1287	0.2923	0.4348	0.5636	0.6859	1.3652	1.4575	1.5182	1.5670	1.6096	1.6484	1.6845	1.7185	1.7508	1.7815
1600 (604.87)	v	0.0161	0.0166	0.0173	0.0185	0.0202	0.0236	0.3415	0.4032	0.4555	0.5031	0.5482	0.5915	0.6336	0.6748	0.7153
	h	72.21	171.69	272.57	376.69	487.60	616.77	1279.4	1358.5	1425.2	1486.9	1546.6	1605.6	1664.3	1723.2	1782.3
	s	0.1286	0.2921	0.4344	0.5631	0.6851	0.8129	1.4312	1.4968	1.5478	1.5916	1.6312	1.6678	1.7022	1.7344	1.7657
1800 (621.02)	v	0.0160	0.0165	0.0173	0.0185	0.0202	0.0235	0.2906	0.3500	0.3988	0.4426	0.4836	0.5229	0.5609	0.5980	0.6343
	h	72.73	172.15	272.95	376.93	487.56	615.58	1261.1	1347.2	1417.1	1480.6	1541.1	1601.2	1660.7	1720.1	1779.7
	s	0.1284	0.2918	0.4341	0.5626	0.6843	0.8109	1.4054	1.4768	1.5302	1.5753	1.6156	1.6528	1.6876	1.7204	1.7516
2000 (635.80)	v	0.0160	0.0165	0.0173	0.0184	0.0201	0.0233	0.2488	0.3072	0.3534	0.3942	0.4320	0.4680	0.5027	0.5365	0.5695
	h	73.26	172.60	273.32	377.19	487.53	614.48	1240.9	1353.4	1408.7	1474.1	1536.2	1596.9	1657.0	1717.0	1777.1
	s	0.1283	0.2916	0.4337	0.5621	0.6834	0.8091	1.3794	1.4578	1.5138	1.5603	1.6014	1.6391	1.6743	1.7075	1.7389
2500 (668.11)	v	0.0160	0.0165	0.0173	0.0184	0.0200	0.0230	0.1681	0.2293	0.2712	0.3068	0.3390	0.3692	0.3980	0.4259	0.4529
	h	74.57	173.74	274.27	377.82	487.50	612.08	1176.7	1303.4	1386.7	1457.5	1522.9	1585.9	1647.8	1709.2	1770.4
	s	0.1280	0.2910	0.4329	0.5609	0.6815	0.8048	1.3076	1.4129	1.4766	1.5269	1.5703	1.6094	1.6456	1.6796	1.7116
3000 (695.33)	v	0.0160	0.0165	0.0172	0.0183	0.0200	0.0228	0.0982	0.1759	0.2161	0.2484	0.2770	0.3033	0.3282	0.3522	0.3753
	h	75.88	174.88	275.22	378.47	487.52	610.08	1060.5	1267.0	1363.2	1440.2	1509.4	1574.8	1638.5	1701.4	1761.8
	s	0.1277	0.2904	0.4320	0.5597	0.6796	0.8009	1.1966	1.3692	1.4429	1.4976	1.5434	1.5841	1.6214	1.6561	1.6888
3200 (705.08)	v	0.0160	0.0165	0.0172	0.0183	0.0199	0.0227	0.0335	0.1588	0.1987	0.2301	0.2576	0.2827	0.3065	0.3291	0.3510
	h	76.4	175.3	275.6	378.7	487.5	609.4	800.8	1250.9	1353.4	1433.1	1503.8	1570.3	1634.8	1698.3	1761.2
	s	0.1276	0.2902	0.4317	0.5592	0.6788	0.7994	0.9708	1.3515	1.4300	1.4866	1.5335	1.5749	1.6126	1.6477	1.6806
3500	v	0.0159	0.0164	0.0172	0.0183	0.0198	0.0225	0.0307	0.1364	0.1764	0.2066	0.2326	0.2563	0.2784	0.2995	0.3198
	h	77.2	176.0	276.2	379.1	487.6	608.4	779.4	1224.6	1338.2	1422.2	1495.5	1563.3	1629.2	1693.6	1757.2
	s	0.1274	0.2899	0.4312	0.5585	0.6777	0.7973	0.9508	1.3242	1.4112	1.4709	1.5194	1.5618	1.6002	1.6358	1.6691
4000	v	0.0159	0.0164	0.0172	0.0182	0.0198	0.0223	0.0287	0.1052	0.1463	0.1752	0.1994	0.2210	0.2411	0.2601	0.2783
	h	78.5	177.2	277.1	379.8	487.7	606.9	763.0	1174.3	1311.6	1403.6	1481.3	1552.2	1619.8	1685.7	1750.6
	s	0.1271	0.2893	0.4304	0.5573	0.6760	0.7940	0.9343	1.2754	1.3807	1.4461	1.4976	1.5417	1.5812	1.6177	1.6516
5000	v	0.0159	0.0164	0.0171	0.0181	0.0196	0.0219	0.0268	0.0591	0.1038	0.1312	0.1529	0.1718	0.1890	0.2050	0.2203
	h	81.1	179.5	279.1	381.2	488.1	604.6	746.0	1042.9	1252.9	1364.6	1452.1	1529.1	1600.9	1670.0	1737.4
	s	0.1265	0.2881	0.4287	0.5550	0.6726	0.7880	0.9153	1.1593	1.3207	1.4001	1.4582	1.5061	1.5481	1.5863	1.6216
6000	v	0.0159	0.0163	0.0170	0.0180	0.0195	0.0216	0.0256	0.0397	0.0757	0.1020	0.1221	0.1391	0.1544	0.1684	0.1817
	h	83.7	181.7	281.0	382.7	488.6	602.9	736.1	945.1	1188.8	1323.6	1422.3	1505.9	1582.0	1654.2	1724.2
	s	0.1258	0.2870	0.4271	0.5528	0.6693	0.7826	0.9026	1.0176	1.2615	1.3574	1.4229	1.4748	1.5194	1.5593	1.5962
7000	v	0.0159	0.0163	0.0170	0.0180	0.0193	0.0215	0.0248	0.0334	0.0573	0.0816	0.1004	0.1160	0.1298	0.1424	0.1542
	h	86.2	184.4	283.0	384.2	489.3	601.7	729.3	901.8	1124.9	1281.7	1392.2	1482.6	1563.1	1638.6	1711.1
	s	0.1252	0.2859	0.4256	0.5507	0.6663	0.7777	0.8926	1.0350	1.2055	1.3171	1.3904	1.4466	1.4938	1.5355	1.5735

Source: Abstracted from *Thermodynamic and Transport Properties of Steam,* © 1967 by the American Society of Mechanical Engineers.

factor in the heat balance. Any water used in blowdown to remove impurities from the boiler water must also be included as lost heat.

8.1.1.3 Heat gains.   A heat balance equates the energy entering a system to the energy leaving the system. To estimate the energy entering a system, it is necessary, first, to determine the amount and form of all energy entering the boiler and, second, to express energy inputs in the same units as outputs, namely Btu/h.

The greatest source of energy input to a boiler is (except for some waste heat boilers) the boiler fuel itself. The energy per pound of fuel is usually one of the criteria for using the given fuel and so is readily available. Fuel moisture subtracts from the amount of energy available from the fuel, and this, along with other calculations for the fuel energy, was explained in Chapter 7.

Other sources of energy in the boiler system include combustion air, returned condensate, and boiler makeup water. The enthalpy in each of these sources is used as the measure of the energy it contributes to the system. The enthalpy of combustion air depends on its moisture content and its temperature. For air containing .013 lb water vapor/lb air (the moisture content at 80°F and 60% relative humidity), the enthalpy at near atmospheric pressure is given in Table 8-3. (For a different humidity, the enthalpy of the moisture can be calculated by integrating the formula for the specific heat of water vapor give in Reference 3.)

The enthalpy of returned condensate can be determined from Table 8-1 from

TABLE 8.3 ENTHALPY OF AIR UNDER ATMOSPHERIC PRESSURE (.013 LB MOISTURE/LB DRY AIR)

Temperature (°F)	Enthalpy (Btu/lb)	Temperature (°F)	Enthalpy (Btu/lb)
0	109.9	950	351.3
50	122.2	1000	364.8
100	134.4	1050	378.3
150	146.8	1100	391.9
200	159.1	1150	405.6
250	171.4	1200	419.3
300	183.8	1250	433.2
350	196.3	1300	447.1
400	208.8	1350	461.0
450	221.4	1400	475.1
500	234.0	1450	489.2
550	246.7	1500	503.3
600	259.5	1550	517.5
650	272.4	1600	531.8
700	285.3	1650	546.2
750	298.3	1700	560.5
800	311.5	1750	575.0
850	324.7	1800	589.5
900	338.0	1850	604.0
		1900	618.6

Source: Specific heat of water vapor taken from Reference 3. Enthalpy of dry air taken from Table 1 in *Gas Tables* by Joseph H. Keenan and Joseph Kaye, © 1948 by John Wiley & Sons, New York.

the temperature or pressure in the condensate line, and the enthalpy of the boiler makeup water, per pound, is the difference in °F between its temperature and 32°F. With these data, it is now possible to present an example showing how an energy balance is calculated and used in the economic evaluation of a proposed boiler improvement.

8.1.1.4 Example.   Consider a boiler with the data shown in Figure 8–1. This is a preliminary input-output diagram. The input and output quantities are expressed in common units so that a preliminary mass balance is easily performed comparing the

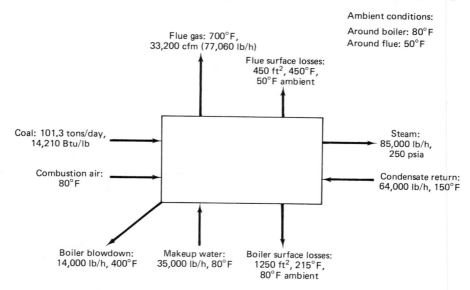

Figure 8–1   Flow balance for boiler.

input quantities to the output quantities and correcting any discrepancy. (A discrepancy of 1–2% is considered acceptable.) The fuel is a Virginia coal, with the ultimate analysis shown in Table 8–4.

TABLE 8.4   FUEL ANALYSES FOR SAMPLE PROBLEM

Ultimate analysis		Flue gas analysis (dry)	
Constituent	% (weight)	Constituent	% (weight)
C	80.3%	CO_2	25.4%
H_2	4.9	SO_2	.1
S	.6	N_2	71.7
N_2	1.7	O_2	2.8
O_2	4.2	(15% excess air)	
H_2O	5.5		
Ash	2.8		

With these data, it is possible to calculate the amount of each energy flow in the boiler and to thereby choose the best place to look for possible energy cost savings. First, calculate the heat losses. They are as follows:

Radiative and Convective Losses (from Equations 8-1 and 8-2). From boiler:

$$radiative\ loss = 1250 \times .1714 \times 10^{-8} \times [(215 + 460)^4 - (460 + 80)^4]$$
$$= 111.4 \times 10^{-8} \times (2.076 \times 10^{11} - 8.503 \times 10^{10})$$
$$= 136,600\ Btu/h$$

$$convective\ loss = 1250 \times .18 \times (215 - 80)^{4/3}$$
$$= 155,800\ Btu/h$$

From flue:

$$radiative\ loss = 450 \times .1714 \times 10^{-8} \times [(450 + 460)^4 - (50 + 460)^4]$$
$$= 476,700\ Btu/h$$

$$convective\ loss = 450 \times .18 \times (450 - 50)^{4/3}$$
$$= 239,700\ Btu/h$$

Therefore, the total radiative and convective losses are approximately 1,009,000 Btu/h.

Heat in Steam.   The energy in the steam is calculated from the enthalpy given in Table 8-1. For 250-psig steam, the enthalpy is 1201.7 Btu/lb. Thus the energy carried from the boiler in steam is 85,000 × 1201.7, or 102,100,000 Btu/h.

Heat in Flue Gas.   To estimate the amount of heat carried away in the flue gas, it is necessary to first estimate the percentage of water vapor in the gas, then to combine this with the flue gas analysis to determine the weight percentages of each gas in the flue gas, and finally to use the specific heat of each gas to estimate the enthalpy increase per cubic foot of escaping flue gas. (Note that this enthalpy is the *increase* from the enthalpy of the entering combustion air. To avoid counting this initial enthalpy twice, the enthalpy of the entering combustion air is counted as zero.) The details of this procedure are outlined in Figure 8.2.

The flue-borne products of each pound of fuel weigh an amount equal to the combustion air + burned constituents + moisture, or, in this example, 12.278 lb

Constituent and lb/lb fuel		lb O_2/lb constituent	Total lb O_2
C:	.803	2.264	1,818
H_2:	.048	7.937	.389
S:	.006	.998	.006
	.857 lb	Total:	2.213
		− O_2 in fuel:	.042
		O_2 needed per lb fuel:	2.171 lb

Figure 8-2   Calculation of heat in flue gas for sample problem.

Air needed = 2.171 lb O_2/lb fuel $\times$ 100/23.15 lb air/lb O_2

　　　　　 = 9.378 lb air/lb fuel

Air used　 = 9.378 $\times$ (1 + excess air percentage)

　　　　　 = 9.378 $\times$ 1.15

　　　　　 = 10.785 lb air/lb fuel

Moisture in flue gas = moisture from hydrogen in fuel
　　　　　　　　　　　 + moisture in fuel + moisture in air
　　　　　　　　　　　 (assumed .013 lb/lb dry air)

Moisture from hydrogen in fuel = $\dfrac{.049 \text{ lb } H_2/\text{lb fuel}}{2 \text{ lb } H_2/\text{lb} \cdot \text{mole } H_2}$ $\times$ 18 lb/lb $\cdot$ mol H_2O

　　　　　　　　　　　　　　　　 = .441 lb

Moisture in fuel　　　　 = .055 lb/lb fuel (from fuel analysis)

Moisture in air　　　　　 = .013 lb H_2O/lb air $\times$ 10.785 lb air/lb fuel

　　　　　　　　　　　　 = .140 lb/lb fuel

Thus total moisture in flue gas = .636 lb/lb fuel

　　　　　　　　　　　　　　　 = $\dfrac{.636}{10.785 + .857}$

　　　　　　　　　　　　　　　 = .0546 lb/lb flue gas

Composition of wet flue gas:

Component	lb/lb dry gas	lb/lb wet gas	/	Molecular weight	=	lb $\cdot$ mol/lb wet gas
CO_2	.254	.241		44		.00548
N_2	.717	.680		28		.02429
O_2	.028	.027		32		.00084
SO_2	.001	.001		64		.00002
H_2O	.0546	.052		18		.00289

Component	lb $\cdot$ mol/lb wet gas	$\times$	Specific heat[a] (Btu/lb $\cdot$ mol, °F)	$\times$	700–80	=	Btu/h
CO_2	.00548		10.4		620		35.3
N_2	.02429		7.1		620		106.9
O_2	.00084		7.3		620		3.7
SO_2	.00002		10.9		620		.1
H_2O (sensible)	.00289 $\overline{.03352}$ lb $\cdot$ mol		8.3		620		14.9 $\overline{160.9}$

Total heat = sensible heat + latent heat

　　　　　 = 160.9 Btu + .052 lb H_2O $\times$ 970 Btu/lb

　　　　　 = 211.3 Btu/lb wet flue gas

Total heat/ft^3 = Btu/lb $\times$ lb/lb $\cdot$ mol $\times$ lb $\cdot$ mol/ft^3

　　　　　 = $\dfrac{211.3}{.03352} \times \dfrac{1}{359} \times \dfrac{460 + 80}{460 + 700}$

　　　　　 = 8.17 Btu/ft^3

Figure 8–2　(*Cont*.)

(10.785 + .857 + .636). Since 101.3 tons/day are used and since each pound of wet flue gas takes with it 155.8 Btu, the loss per day from the flue is given by 101.3 × 2000 × 12.278 × 155.8 = 387,600,000 Btu/day, or 16,150,000 Btu/h. Note that this represents a loss of 155.8 × 12.278 = 1910 Btu/lb. Since the heating value of this coal is 14,210 Btu/lb, the flue loss represents a loss of about 13.4%. Note also that the heat carried by 33,200 cfm of flue gas is 33,200 × 60 × 8.17 = 16,275,000 Btu/h, close to that calculated from the fuel measurements alone.

Heat in Boiler Blowdown.   The enthalpy of water at 400°F is 375.1 Btu/lb, and hence the blowdown of 14,000 lb/h represents a loss of 5,251,000 Btu/h.

This procedure gives a way to calculate the heat leaving a system. This loss must be equal to the energy entering the system. Sources of this energy are fuel, combustion air, makeup water, and returned condensate.

Fuel.   As indicated, the amount of energy available from the fuel is the rated fuel heating value multiplied by the amount of fuel used per hour. In this example, the energy input is [(101.3 tons/day)/(24 h/day)] × 2000 lb/ton × 14,210 Btu/lb, or 120 million Btu/h.

Combustion Air.   As shown in Figure 8-1, the temperature of incoming combustion air is 80°F. Since the enthalpy of flue gas was calculated relative to the temperature of the entering combustion air, the relative enthalpy of the combustion air is zero. This convention avoids counting the enthalpy of combustion air both in the input and the output computations.

Makeup Feedwater.   The makeup water represents an energy input of 3000 lb/h × (80 − 32)°F, or 144,000 Btu/h.

Condensate Return.   Condensate returned to the boiler accounts for a substantial amount of energy input. In the preceding example, this input is 64,000 lb/h × (150 − 80) Btu/lb, or 3,360,000 Btu/h.

It is convenient to summarize the input and output energy flows as shown in Figure 8-3. As before, the input and output enthalpies per hour should be within 1-2% of each other.

The heat balance of Figure 8-3 serves several objectives. First, it is a clear way of presenting the energy input and output of a system, whether it is a boiler or an industrial process. This diagram can thus be used to communicate the energy flows, whether the communication is part of the data-checking process or whether it is part of a presentation being made in defense of a particular proposal. Second, preparing the diagram necessitates examining the system in detail in order to find and quantify the energy inputs and outputs. Many of these may not have been evaluated before, and their magnitudes are probably different from what they are thought to be. Third,

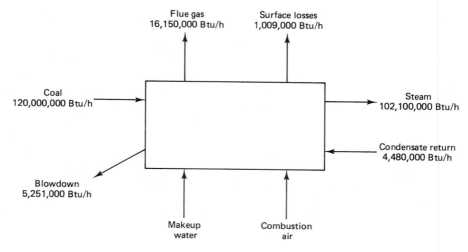

Figure 8-3   Heat balance for boiler.

the balance immediately and visibly suggests ways the system operation may be improved. Some of these possibilities are presented in the problems at the end of this chapter.

8.1.2 Insulation

One of the most common methods for reducing heat loss from boilers is the addition of pipe and surface insulation (see Table 8-5). The choice of insulation type and thickness, together with the economics of insulation, are treated in detail in Chapter 11.

TABLE 8.5 PROPERTIES OF INSULATING MATERIALS[a]

Material	Maximum temperature	Conductivity per inch thickness	Resistance (R) per inch thickness
Glass fiber blanket	1200°F	.26–.54	1.8–3.8
Glass felt, organic bonded	350	.27–.68	1.5–3.7
Asbestos paper (6-ply)	300	.49–.59	1.7–2.0
Molded amosite	1500	.32–.72	1.4–3.1
Calcium silicate	1800	.63–.95	1.1–1.6
Slag or glass pipe insulation	1000	.33–.55	1.8–3.0
Mineral fiber insulating cement with clay binder	1800	.49–.85	1.2–2.0

[a]Higher conductances are associated with higher temperatures for the materials shown. For exact values for a particular manufacturer, consult vendors.

Source: Reference 5.

8.2 WASTE HEAT RECOVERY

Waste heat is heat that is given to the atmosphere or to some other heat sink without providing appreciable benefit to the user. Examples of such heat loss include flue gases, boiler blowdown exhausted to the air, heated air from vents exhausted directly to the environment, and heat lost from pipes that pass through a space where heating is not needed. (Note that where these pipes provide protection against freezing, the heat radiated from the pipes is not waste heat.) Waste heat can be used to generate steam, as a source of energy for turbines, or as a source of free heat for incoming fluid streams. Any of these can be used to improve energy utilization in a plant, whether the source of the waste heat is a boiler, an industrial process, or an HVAC system. Benefits from waste heat recovery often include fuel savings and lower capital cost of heating and cooling equipment. Other benefits can include increased production capacity and, under some circumstances, revenue from the sales of recovered heat or energy [4].

8.2.1 Analyzing for Waste Heat Recovery Potential

Waste heat sources and their uses can be conveniently categorized by the temperature at which the heat is exhausted, as shown in Table 8-6 [4].

TABLE 8.6 SOURCES OF WASTE HEAT

Temperature	Source	Use
High (1200 +)	Exhausts from direct-fired industrial processes: Cement kiln (dry), 1150–1350°F Steel heating furnaces, 1700–1900°F Glass melting furnaces, 1800–2800°F Solid waste incinerators, 1200–1800°F Fume incinerators, 1200–2600°F	Cogeneration
Medium (450–1200)	Exhausts: Steam boiler, 450–900°F Gas turbine, 700–1000°F Reciprocating engine, 450–1100°F Heat-treating furnaces, 800–1200°F Drying and baking ovens, 450–1100°F Annealing furnace cooling systems, 800–1200°F	Steam generation
Low (90–450)	Process steam condensate, 130–190°F Cooling water from: Furnace doors, 90–130°F Bearings, 90–190°F Welding machines, 90–190°F Air compressors, 80–120°F Internal combustion engines, 150–250°F Hot-processed liquids, 90–450°F Hot-processed solids, 200–450°F	Supplemental heating Preheating

Source: Reference 4.

The use of waste heat to power a turbine or pump should be considered if there is a need for turbine work or pumping and if the energy available is enough to justify the cost involved. If the economics justify the production of electricity, cogeneration, discussed in Section 8.5, may be a possibility. The use of waste heat from one source to heat another fluid stream should be considered if (1) the waste heat source is close enough to the heat sink that the fluid temperature will still be high enough to be useful even after taking into account all heat lost in transporting the fluid from source to sink, (2) using waste heat from the source will not create problems at the source, and (3) the transfer of heat from the source to the sink is technically feasible.

The first step in the analysis of an industrial process for possible waste heat recovery is the collection of data sufficient to describe the process with a heat balance such as that of Figure 8-3. The next step is to find and quantify all point sources of heat use or exhaust, with the annual Btu and mean temperature for each, and to show this information on an outline of the facility. Another way to summarize these data is shown in Figure 8-4. The most promising candidates for heat recovery are then examined in detail with the kind of data shown in Figure 8.5. If at all possible, use the waste heat to improve the process at the heat source to avoid transportation losses and to keep processes as independent from one another as possible.

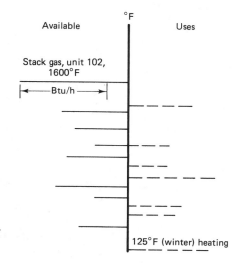

Figure 8-4   Source-sink diagram for waste heat recovery. (*State of New Jersey Energy Audit Forms*, 1979. Reprinted with permission of the State of New Jersey Dept. of Energy)

8.2.2 The Economics of Waste Heat Recovery

The benefits from waste heat recovery can be substantial, and it is important that the benefits included in the economic analysis should be as complete as possible. Table 8-7 gives one list—this should be amplified by consideration of the virtues of each specific project. At the same time, the analysis should give complete details of the costs involved and the amount by which these costs are reduced by tax benefits. Table 8-8 gives such a list.

Source data

   Location: _____

   Fluid used as a heat source: _____

   Flow rate: _____ lb/h

   Present temperature of fluid: _____ °F

   Present pressure of fluid: _____ psig

   Dew point of fluid: _____ °F

   Minimum allowable temperature: _____ °F

Sink data

   Location: _____

   Fluid to be heated: _____

   Flow rate: _____ lb/h

   Present temperature of fluid: _____ °F

   Maximum allowable temperature of fluid: _____ °F

   Present pressure of fluid: _____ psig

   Maximum containable pressure of fluid: _____ psig

Heat transportation path data

   Horizontal distance: _____ ft

   Vertical distance: _____ ft

   Mass pumped: _____ lb/h

   Pumping cost: $ _____ /h

   Estimated heat loss per foot: _____ Btu/h

   Heat transported: _____ Btu/h

   Temperature drop: _____ °F

   Pressure drop: _____ psig

General data

   Building location: _____

   Heating cost, for planning purposes: $ _____ /Btu

   Electricity cost, for planning purposes: $ _____ /kWh

   Production: _____ (amount and units)/year

   Production cost savings: $ _____ /unit output

Figure 8-5   Data needed to do an in-depth analysis of a particular waste heat recovery measure.

TABLE 8.7   ECONOMIC BENEFITS FROM WASTE HEAT RECOVERY

Benefits	How quantified
Reduced fuel costs	$/Btu × Btu/h × operating h/year
Increased production capacity	Marginal profit/unit × increase in capacity × no. h increased capacity will be used each year
Reduced overhaul costs	Cost per overhaul/mean time between overhauls
Reduced regular maintenance costs	Cost of maintenance per h × no. h of maintenance per operating h
Lower capital costs of furnaces and heating and cooling equipment	Vendors' quotations
Sales of energy to utilities	*Firm* contracts with utilities, or regulations of state public service commission
Reduced pollution abatement equipment costs	Vendors' quotations

TABLE 8.8 COSTS AND COST REDUCTIONS TO BE CONSIDERED IN WASTE HEAT RECOVERY

Cost	Source of information
Capital cost of equipment	Vendors
Energy tax credits	Internal Revenue Service
Capital investment tax credits	Internal Revenue Service
State and local tax credits	State and local tax information sources
Increased property taxes	State and local tax information sources
Engineering design costs	Local consulting engineers (plan on 2–5% of total project costs)
Installation costs	Time and equipment cost estimates
Cost of production downtime during installation	In-house production, time, and cost estimates
Operation and maintenance costs of new equipment	Vendors

Source: Taken in part from Reference 4.

8.2.3 Waste Heat Recovery Equipment

The equipment for waste heat recovery depends on the fluid temperature at the source, the intended use for the waste heat, and the distance the heated fluid (if any) must be transported. The most important equipment types and their uses are shown in Table 8–9.

TABLE 8.9 WASTE HEAT RECOVERY EQUIPMENT

Source fluid	Use	Equipment
Exhaust gases over 450°F	Preheat combustion air:	
	Boilers	Air preheaters
	Furnaces	Recuperators
	Ovens	Recuperators
	Gas turbines	Regenerators
Exhaust gases under 450°F	Preheat boiler feedwater or makeup water	Economizers
Condenser exhaust gases	Preheat liquid or solid feedstocks	Heat exchangers
Any waste heat source	Generate steam, power, or electricity	Waste heat boilers
Any waste heat source	Transfer heat	Run-around coils, heat wheels
Any waste heat source	Cooling	Absorption chillers

Source: Reference 4.

8.2.3.1 Recuperators.
A recuperator is a heat transfer device that passes gas to be heated through tubes that are surrounded by the gas that has excess heat. The heat is transferred from the hot gas to the tubes and through the tube walls to the cool gas inside the tubes. Examples of two types of recuperators are shown in Figures 8–6 and 8–7.

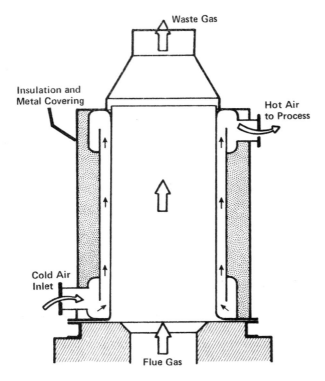

Figure 8-6 Radiation recuperator. (From Kenneth G. Kreider and Michael B. McNeil, *Waste Heat Management Guidebook*, National Bureau of Standards Handbook 121, 1977. Courtesy of the U.S. Department of Commerce.)

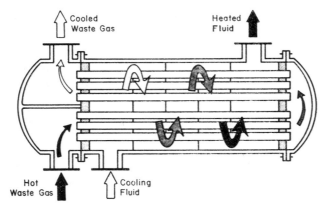

Figure 8-7 Convective-type recuperator. (From Kenneth G. Kreider and Michael B. McNeil, *Waste Heat Management Guidebook*, National Bureau of Standards Handbook 121, 1977. Courtesy of the U.S. Department of Commerce.)

8.2.3.2 Heat wheels. A heat wheel is a large porous wheel that rotates between two adjacent ducts, one of which contains a hot gas and the other a cooler gas. In carrying heat from the hot gas to the cool gas, the heat wheel either captures some of the escaping heat in the hot gas, if that is desired, or cools the hot incoming gas, if that is desired. If the escaping gas is contaminated, it is possible to build a purge section into the wheel whereby the contaminants are flushed from the wheel after they have given up a significant part of their heat. These wheels, illustrated in Figures 8-8 and 8-9,

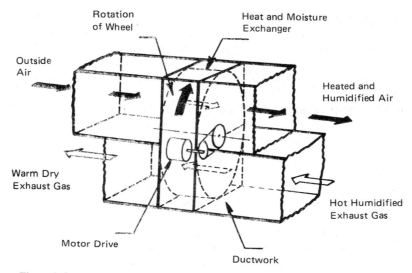

Figure 8-8 Simple heat wheel. (Used with permission of the American Society of Heating, Refrigerating, and Air Conditioning Engineers, Inc., Atlanta, Ga.)

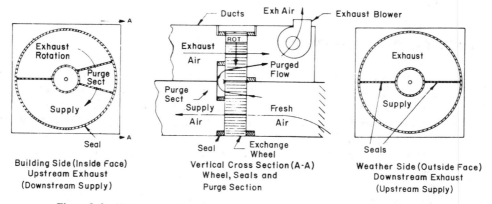

Figure 8-9 Heat wheel with purge section. (From Kenneth G. Kreider and Michael B. McNeil, *Waste Heat Management Guidebook*, National Bureau of Standards Handbook 121, 1977. Courtesy of the U.S. Department of Commerce.)

have high heat transfer efficiencies and can be used to significantly reduce either heating or cooling loads. It should be noted, moreover, that they can be designed to transfer latent heat, i.e., the heat in water vapor in the gas stream, as well as the usual sensible heat.

8.2.3.3 Air preheaters.   In an air preheater, hot gas through a series of closed channels gives off heat to cooler gas in adjacent channels. Through the use of this kind of equipment, hot flue gas can be used to preheat combustion air, reducing the amount of heat that must be supplied by fuel.

8.2.3.4 Run-around coils. A run-around coil heat exchanger consists of two coils and the piping connecting them. The heat exchange fluid picks up heat in one coil from the source, is pumped to the sink, and gives up its heat there. This heat transfer method makes possible waste heat recovery when the source and sink are somewhat separated, but it does not allow for the transfer of latent heat.

8.2.3.5 Finned-tube heat exchanger. This is the usual kind of heat exchanger found in baseboard heating—a tube surrounded by perpendicular fins. The fins help to transfer heat from the tube to the surrounding air by enlarging the heat transfer surface area. If the surrounding fluid has a higher temperature than the tube, the transfer works in the opposite way and transfers heat from the fluid to the material inside the tube. This type of heat exchanger is very common and is often used on boilers to recover some of the heat that would otherwise be lost in the stack gas.

8.2.3.6 Waste heat boilers. If gas leaving some industrial process is sufficiently hot to vaporize water or some other working fluid, it may be possible to use a waste heat boiler. Such a boiler uses waste heat to create vapor or steam and then generates electricity or work by running the vapor through a turbine, pump, or generator. The customary fluid for this purpose has been steam, but working fluids with lower boiling points are becoming more common.

8.3 BOILER MANAGEMENT

In preceding sections of this chapter we have shown how to determine the amount of heat lost in a boiler and the kind of equipment needed to recover heat being wasted. In Chapter 7 we discussed combustion in boilers with particular attention to combustion air. In addition to monitoring combustion air closely, good boiler management requires close attention to the boiler control system and to balancing the load economically between boilers. There are also other opportunities for saving money and increasing production with boilers, and they are discussed in this section.

8.3.1 Boiler Types

Boilers can be categorized into several types: packaged boilers, field-erected boilers, and fluidized-bed boilers. Packaged boilers are delivered to a customer with all the controls attached, ready for installation upon arrival. They are manufactured, low-capacity (350,000–35,000,000 Btu/h at 300 psig maximum) units and are designed to be operated with a minimum of skill and attention. They can be either a fire tube type (with water surrounding the tubes), as shown in Figure 8–10, or a water tube type.

In contrast to the packaged boiler, the field-erected boiler is a large-capacity, custom-built unit with specialized controls and usually with the requirement of specially trained operators. The water in these boilers is in the tubes that make up the wall of the combustion chamber and in the economizers and other boiler components

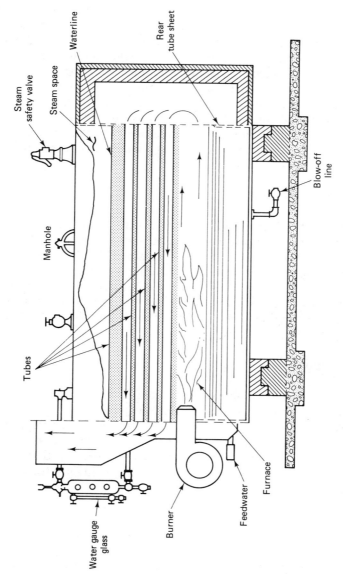

Figure 8-10 Packaged Scotch boiler. (Courtesy of American Boiler Manufacturer's Association.)

169

designed to contribute directly to steam distribution. Such boilers are complicated and are discussed in detail in References 6 and 7.

Fluidized-bed boilers are becoming more important in energy management because of their ability to burn many kinds of industrial wastes as fuel with very little air pollution. In most boilers, fuel is either burned in front of burners within the combustion chamber or burned on a traveling grate. In a fluidized-bed boiler, however, air is forced upward through a series of holes into the bottom of a bed of broken limestone, and fuel is inserted on top of the limestone. The air holds the limestone in suspension and provides combustion air for the fuel; the limestone captures sulfur compounds in the fuel and combines with them to form sulfates, solid products that can be removed and used for fertilizer. In addition to solving the problem of sulfur, fluidized-bed boilers also burn at a temperature lower than ash-softening temperatures, resulting in much simpler ash handling than with conventional boilers. Also, soot blowers are not required [8].

8.3.2 Load Management

Boilers differ in efficiencies, and, in a system with several boilers, it makes sense to determine which boilers are the most efficient and to develop a method for allocating the load to them in order of decreasing efficiency. This allocation can be done either manually, by turning boilers down or off during off-peak seasons, or on a real-time basis through computer control. A boiler operates inefficiently with low loads, and it is usually worthwhile to operate one boiler at 90% capacity rather than two boilers at 45% capacity or three at 30%. The automatic controls needed are described in Reference 9, or information can be obtained from vendors.

8.3.3 Components

Before leaving boilers, it should be noted that efficiency can be improved by exchanging burners and other components for more efficient models as they are developed. As with other new technology, the credibility of the manufacturer is very important in estimating the value of any promotional literature. Besides burners, other components that should be examined are those wherein the heat balance indicates that waste heat may be successfully used—as, for example, in units for preheating fuel oil or other fuels.

8.4 STEAM DISTRIBUTION SYSTEMS

Steam distribution systems consist of piping from a boiler to points of use, together with steam traps, condensate return lines, and any pumps needed for condensate. The main function of the steam distribution system is to get the steam to where it is needed and to return the condensate to the boiler, both as efficiently as possible. Energy management affects this system through insulation, detection and repair of steam and condensate leaks, steam trap and condensate pump maintenance, and water treatment.

8.4.1 Insulation

A great deal of energy can be saved by insulating pipes that carry either steam or condensate. This subject is described in detail in Chapter 13.

8.4.2 Steam Leaks

The number of pounds of steam per month through an orifice of area A can be found from Grashof's formula [3, pp. 4-47 and 4-48] as

$$\text{lb/h} = .70 \times .0165 \times 3600 \times A \times P_1^{.97} \tag{8-3}$$

where $.70$ = coefficient of discharge for hole (for a perfectly round hole, this is 1)
$.0165$ = a constant in Grashof's formula
3600 = number of seconds per hour
A = area of hole in in.^{2}
P_1 = pressure inside steam line in psia

With this formula and with Table 8-1, it is possible to develop tables giving steam and Btu losses for any given leak size for a plant whose steam pressure is known. When the cost per million Btu is also known, it is then possible to estimate the cost of steam leaks directly from the size of the leaks. Table 8-10 gives an example of such calculations for a 600-psig system, assuming that the energy cost is $6.00/million Btu.

Another way to estimate steam losses is given by Waterland [10]:

A crude but frequently quite effective way of evaluating steam leaks is based on an arbitrary rating system. Tour a defined plant area at a weather condition or time of day when leaks are quite prominent, making a note of each steam leak and rating it as a wisp, moderate leak, or severe leak. Assign a value of 25 lb/hr for each wisp, 100 lb/hr for a moderate leak, and 500 to 1000 lb/hr for the severe leaks. When more than 20 leaks are evaluated, the total leakage rate determined in this way will usually be within 25% of the actual steam loss.

TABLE 8.10 HEAT AND DOLLAR LOSSES FROM LEAKS OF 600-PSIG STEAM

Hole diameter	Steam loss (lb/h)	Heat loss (Btu/h)	Dollar loss ($/month)
1/16	74	89,000	$390
1/8	296	356,000	1,560
1/4	1183	1,420,000	6,230
3/8	2660	3,200,000	14,030
1/2	4730	5,690,000	24,942

8.4.3 Steam Traps

Anything that reduces heat transfer from steam to the pipe walls can be a source of inefficiency. In particular, air and dissolved gases act as insulators and should be removed, along with condensed steam, as soon as possible within the steam distribution system. This removal is one of the functions of *steam traps*. The main function of steam traps, however, is to drain condensed steam from the steam distribution network so that it can be returned to the boiler.

The maintenance of steam traps is important in energy management because condensate, dissolved air, and certain gases must be removed from steam lines if the lines are to transmit steam. If condensate is not removed, the steam lines become lines carrying water but transmitting little heat. If dissolved air is not removed, the steam carries significantly less heat per pound since the pressure of the steam is reduced by the partial pressure of air. If dissolved CO_2 is not removed, carbonic acid is formed, and this has a corrosive effect upon pipes. Any of these things can happen if steam traps fail closed. If steam traps fail open and they are open to the air, the effect is the same as if there were a large leak in a steam line. Steam trap maintenance can be a very important source of energy cost savings. This topic is covered in detail in Chapter 14 of Reference 1.

8.4.4 Water Treatment

Finally, consider the effect of scaling on heat transfer. The more scale buildup, the less heat is transmitted through pipe walls. Reference 7 gives an example where a layer of $CaSO_4$ only .024 in. thick caused a temperature drop of 362°F in a boiler, leading to an outer tube temperature of 1004°F and to ultimate failure of boiler tubes. This scaling can be prevented by proper *water treatment,* making water treatment one of the essential elements in boiler management. The amount of water treatment needed depends on the hardness of the water and the quantity of water used and is one of the more cogent reasons for reusing returned condensate.

8.5 COGENERATION

Cogeneration is the generation of high-temperature steam and the use of this steam to both generate electricity and to act as a source of process heat. In large utility boilers, it is customary to generate high-pressure steam, to extract work from this steam through a series of steam turbines, to condense the steam through an elaborate cooling system using either water or air as the fluid receiving the waste heat, and to return the condensate to the boiler for reheating into steam. This cooling process is inherently inefficient—no useful work is being done by the waste heat lost in cooling towers or in water-cooled condensers. Cogeneration—the simultaneous generation and use of electricity and process heat from the same prime mover—does away with some of these inefficiencies and gets more useful work from the energy input into the prime mover. Cogeneration has recently become more attractive because any cogenerated electricity that contributes to the total peak deliverable capacity of an electric utility enables the utility to avoid building an equal amount of new capacity, and the utility is obligated to buy such electricity from the cogenerator. (There are strict conditions on this purchase: The company must use some of the electricity it produces, the cogenerated electricity must be controlled and interconnected in such a way that it will not damage the utility network, and it must not force a net loss upon the utility.)

Cogeneration is not a new idea. In process industries where a great deal of heat is needed, many of the facilities were designed with cogeneration in mind. For example, one of the largest producers of magnesium takes the magnesium from magnesium chloride by electrolysis. The magnesium chloride comes from salt brines that have been dried. Drying the salt brines takes a great deal of heat; the electrolysis uses large quantities of electricity. The only way that this process can be carried out economically is to generate the electricity and use the waste heat to drive off the water from the brines. Another similar use is in refineries.

In addition to industrial process use, cogeneration has also been used for district heating, where steam or hot water is supplied to a district by a utility whose primary product is electricity. Such use became less prevalent after World War II, when fuel was cheap and the economies of returned condensate were more attractive than those of wide-area distribution of steam and hot water. More recently, cogeneration is being used in large apartment buildings and in shopping centers as a less expensive source of both heating and electricity.

8.5.1 Cogeneration Cycles

There are three principal ways in which heat and electricity are produced by the same process: topping cycles, bottoming cycles, and heat recovery from diesel engines. In a topping cycle, steam is generated at high temperature and pressure and is used to drive a steam turbine which generates electricity. Steam leaving the turbine is then used for process heat, often through heat exchangers. In a bottoming cycle, the high-temperature fluid from the prime mover (a boiler of some kind) is first used for process heat, with the waste heat from the industrial process used to generate electricity, generally through the use of some fluid with a low boiling point. The technology of bottoming cycles is newer than that of topping cycles, and the equipment is heavier because of the need to move more fluid to get the same amount of heat or electrical energy. The prime mover in a diesel cycle consists of a diesel engine. This engine drives a generator, and the cooling water and cooling oil are used as sources of heat.

8.5.2 Conditions for Successful Application

Cogeneration offers the promise of cheap electricity, with the electrical price under the control of the plant owner rather than of a utility company. But this promise may well be an illusion unless at least five conditions are met.

Condition 1.   The need for heat must coincide with the need for electricity unless there are facilities for substantial heat or electrical energy storage. For example, one major university uses 400°F water for building heat and has an average electrical demand of 1.15 Mw with peak demands of 2.3 Mw. Cogeneration seemed feasible in this situation until it was noted that the highest electrical usage came from air conditioning in the summer at the same time that the demand for heat was a minimum. This imbalance might even then have been remedied if the campus used air conditioning based on a central chilled water facility with absorbing chillers; unfortunately, such a facility would have involved a major retrofit and would have been prohibitively expensive. Another problem with this cogeneration proposal was that the university would have had a surplus of electricity at the same time as that of the local electric utility, thus driving down the price the utility would have been willing to pay for the electricity. All these factors were studied by consulting engineers; together they amounted to a convincing argument against cogeneration in this case.

Condition 2.   Unless the economics are overwhelming, retrofitting should not be considered. Bottoming cycle equipment, in particular, is usually very heavy and can require substantial renovating of building foundations. To define these and other installation problems, it is a good idea to get a detailed feasibility study from a reputable engineering firm with experience in bottoming cycles. It is also obvious but notable that a facility designed to produce electricity in addition to process heat now being provided (or vice versa) will necessarily be larger than the present facility. This increase in scale is usually expensive. In addition, it may create other problems. The authors have had experience with a lumber products operation where the boiler was operated completely on hogged fuel—chopped tree wastes. Changing this unit into a cogeneration facility would have required more energy than was available from the hogged fuel, necessitating either the purchase of gas or the development of some new fuel source. These mill operators were not enthusiastic about cogeneration.

Condition 3.   Any sale of electricity planned into the economics must be based on legally binding documents, and the conditions under which the electricity can be sold, particularly to utilities, must be known and their economic consequences evaluated. An electric utility is under no obligation to buy electricity in any way that would result in a net loss. In particular, an electric company is not obligated to pay premium rates for electricity delivered when the company has a surplus, and the company is not obligated to tie into a less controlled electric source having a phase or frequency difference that could harm the electrical system. The cost and constraints imposed by such controls should be considered and carefully evaluated—they can be great enough to discourage this particular use of electricity in favor of using all the electricity in the cogenerating plant.

Condition 4.   The source of fuel for the prime mover should be considered carefully. When electricity is to be sold to a utility, the price is based in part on the reliability of the cogenerating source, and this places greater burdens on the cogenerator. If the source of fuel is industrial waste of some sort, this may have to be supplemented with purchased fuel at an unacceptable cost.

Condition 5.   The cogenerator must be prepared to purchase or provide backup heat and/or electricity for occasions when the cogeneration unit is out of service. This is an important economic consideration.

8.6 SUMMARY

In this chapter we have presented the concept of a heat balance and have given useful areas for its application in boilers and waste heat utilization. In addition, the general outlines of cogeneration have been presented. It is left for the readers to innovate from here.

REFERENCES

1. Philip S. Schmidt, "Steam and Condensate Systems," in *Energy Management Handbook,* Wayne C. Turner, Senior ed., Wiley-Interscience, New York, 1982, Chap. 5.
2. Frank Kreith, *Principles of Heat Transfer,* International Textbook Company, New York, 1962.
3. Theodore Baumeister, Eugene A. Avallone, and Theodore Baumeister, III, *Marks' Standard Handbook for Mechanical Engineers,* McGraw-Hill, New York, 1978.
4. Kenneth G. Kreider and Michael B. McNeil, *Waste Heat Management Guidebook,* National Bureau of Standards Handbook 121, U.S. Department of Commerce, Washington, D.C., 1977.
5. *ASHRAE Handbook: 1977 Fundamentals,* American Society for Heating, Refrigerating, and Air Conditioning Engineers, Atlanta, Ga., 1977.
6. Joseph G. Singer, *Combustion,* Combustion Engineering, Inc., Windsor, Conn., 1981.
7. *Steam/Its Generation and Use,* Babcock and Wilcox, New York, 1978.
8. *Heat Engineering,* Foster Wheeler Corporation, Livingston, N.J. July–Sept. 1982.
9. Francis G. Shinskey, *Energy Conservation Through Control,* Academic Press, New York, 1978.
10. Alfred F. Waterland, "Energy Auditing: A Systematic Search for Energy Savings Opportunities," in *Energy Management Handbook,* Wayne C. Turner, Senior ed., Wiley-Interscience, New York, 1982, Chap. 3.

BIBLIOGRAPHY

Steam Conservation Guidelines for Condensate Drainage, Armstrong Machine Works, Three Rivers, Mich., 1976.

TURNER, W. C., *Waste Heat Recovery Module,* Oklahoma State University Industrial Energy Conservation and Management Program, Stillwater, Okla., 1980.

QUESTIONS

8.1. Conceivably, most of the heat in the flue gas should be recoverable. What are the practical considerations that limit the amount of heat recovery?

8.2. What are some ways suggested by Figure 8–3 for improving the efficiency of that boiler system?

8.3. In Problem 8.4 in the following section, it is unlikely that the pressure of the entire steam system will be 350 psig because of steam drops. Does this make the use of Grashof's formula invalid if 350 psig is used for P_1?

PROBLEMS

8.1. A tour of a 600-psi steam distribution system shows 50 wisps (estimated at 25 lb/h), 10 moderate leaks (estimated at 100 lb/h), and 2 leaks estimated at 750 lb each [10]. The boiler efficiency is 85%, the ambient temperature is 75°F, and the fuel is coal, $65.00/ton, at 14,500 Btu/lb. The steam system operates continuously throughout the year. How much do these leaks cost per year in lost fuel?

8.2. Superheated steam enters a heat exchanger at 1400°F and 500 psia and leaves as water at 300°F and 120 psia. How much heat is exchanged per pound of entering steam?

8.3. What would be the amount of potential annual savings in the example of Section 8.1.1.4 if the amount of boiler blowdown could be decreased to an average rate of 3000 lb/h, assuming that it remained at 400°F? How much additional heat would be available from the 3000 lb/h of blowdown water for use in heating the incoming makeup water? Calculate the combined cost saving of these two measures using a fuel cost of $65.00/ton of 14,200-Btu/lb coal.

8.4. Suppose that you are preparing to estimate the cost of steam leaks in a 350-psig steam system. The source of the steam is 14,200-Btu/lb coal at $70.00/ton, and the efficiency of the boiler plant is 70%. Hole diameters are classified as $\frac{1}{16}$, $\frac{1}{8}$, $\frac{1}{4}$, $\frac{3}{8}$, and $\frac{1}{2}$ in. Develop a table showing the size of the orifice, the number of pounds of steam lost per hour, the cost per month, and the cost for an average heating season of 7 months.

Chapter 9

....

Control Systems
and Computers

9.0 INTRODUCTION

Energy utilization can be *controlled* in order to reduce costs and maximize profits. Sometimes the control required is as simple as turning off a switch, but sometimes automated controls ranging from simple clocks to sophisticated computers are required. Our philosophy is that the control should be as simple and reliable as possible. Consequently, this chapter starts with manual controls and proceeds (in increasing complexity) through timers, programmable controllers, and computers.

As one proceeds through this hierarchy of controls, capital is traded for automation and complexity. That is, the automated controls are expensive, but they can do more. Picking the proper type of control is often a difficult task, but we will explore this decision process.

Computers can also help the energy manager in the analysis of proposed and present energy systems. Some excellent large-scale computer simulation programs have been written that enable the analyst to try alternative scenarios of energy equipment and control, so in the last part of this chapter we discuss these computer programs and their optimal use. BLAST 3.0 and DOE 2.1A are the two analyzed in depth, but several others are mentioned.

9.1 TYPES OF CONTROLS

As mentioned in Section 9.0, the different types of controls will be presented in order of increasing complexity and cost. The philosophy of the presentation is that each level of control can perform the functions discussed in that section as well as all those

177

discussed in the simpler systems. For example, the functions discussed in the section on timers can be done very well by a timer or any of the succeeding types of controls (programmable controllers, microprocessors, and large computers) but not by a manual system.

9.1.1 Manual Systems

The idea here is to turn it off when it's not needed. For example, lights are often left on at night. They should be turned off whenever possible. (Often a small series of lights is left on for security purposes.)

One of the best opportunities for manual control exists in the area of exhaust and makeup air fans. Often these fans are located at the top of a tall ceiling where the fan can be left running, but where its running is undetectable without close scrutiny. The savings for turning off exhaust fans is twofold. First, electricity is no longer required to turn the fan motor, and, second, conditioned air is no longer being exhausted.

Suppose, for example, that a fan is exhausting 10,000 ft³/min from a welding area. The fan is run by a 5-hp motor and is needed for two shifts (8:00 a.m. to 12:00 midnight) 5 days/week. Previously, the fan has been left running all night and on weekends. If the space is not air-conditioned and is heated by a gas furnace that is 80% efficient, what is the savings for turning the fan off at night and on the weekends? Gas costs $5.00/million Btu and electricity, $.08/kwh (there will be no demand savings since peaking probably does not occur at night). Assume the outside temperature averages 30°F for the hours the fan can be shut off. (This would have to be determined through analyses, discussed in Chapter 6.)

1. Electricity savings:

$$(5 \text{ hp}) \left(\frac{.746 \text{ kw}}{\text{hp}} \right) \left[\left(\frac{5 \text{ days}}{\text{week}} \right) \left(\frac{8 \text{ h}}{\text{day}} \right) + \left(\frac{2 \text{ days}}{\text{week}} \right) \left(\frac{24 \text{ h}}{\text{day}} \right) \right] \left(\frac{52 \text{ weeks}}{\text{year}} \right)$$

$$\times \left(\frac{(\$.08)}{\text{kwh}} \right)$$

$$= \$1365.48$$

2. Heating savings: (see page 116)

$$\left(\frac{10,000 \text{ ft}^3}{\text{min}} \right) \left(\frac{60 \text{ min}}{\text{h}} \right) \left[\left(\frac{5 \text{ days}}{\text{week}} \right) \left(\frac{8 \text{ h}}{\text{day}} \right) + \left(\frac{2 \text{ days}}{\text{week}} \right) \left(\frac{24 \text{ h}}{\text{day}} \right) \right]$$

$$\times \left(\frac{52 \text{ weeks}}{\text{year}} \right) \left(\frac{.078 \text{ lb}}{\text{ft}^3} \right) \left(\frac{.248 \text{ Btu}}{\text{lb} \cdot {}^\circ\text{F}} \right) (65 {}^\circ\text{F} - 30 {}^\circ\text{F}) \left(\frac{\$5.00}{10^6 \text{ Btu}} \right) \left(\frac{1}{.8} \right)$$

$$= \$11,618.01$$

where the density of air $= \dfrac{.078 \text{ lb}}{\text{ft}^3}$ and the specific heat of air $= \dfrac{.248 \text{ Btu}}{\text{lb} \cdot {}^\circ\text{F.}}$

3. Total savings $= \$12{,}983.49$.

As can be seen, turning equipment off when not in use can lead to dramatic savings. As another example, a large office complex made a detailed study of building utilization and found that only a few tenants worked at nights or on the weekends. By making provisions for these few, the office complex was able to reduce lighting and space conditioning, saving about one third of their energy bill.

The preceding example mentioned night setback. For most industrial plants and many office buildings, this offers significant savings with little to no capital expenditure. (Timers can be purchased and installed to perform night setback, but since this section deals with manual controls, we will assume setback is done manually.)

The calculation procedure for savings is relatively simple and involves heat loss calculations during the hours of setback. Bin data on outside temperature and inside thermostat settings are required. One simply calculates heat losses with the old thermostat setting and again with the revised setting. The difference is the heating savings in Btu.

To simplify this or at least to give an approximation, a nomograph is given in Figure 9–1. The following example shows how to run the calculation.

A manufacturing company of 100,000 ft² is located in an area where heating demands are 4000°F days. The company keeps its thermostats set at 70°F all the time even though it works only one shift. If the company pays $4.50/10⁶ Btu for its natural gas and the heaters are 75% efficient, what would the savings be for turning the thermostats back to 55°F at night when the building isn't occupied? Presently, the company figures it consumes 240×10^3 Btu/ft² of gas for heating. (Normally, this can be estimated from gas bills.)

As can be seen from Figure 9–1 (follow the heavy black lines), the savings are approximately 125×10^3 Btu/ft². Total savings then are

$$\text{fuel savings in Btu} = (125 \times 10^3 \text{ Btu/ft}^2)(100{,}000 \text{ ft}^2) \times \frac{1}{.75}$$
$$= 16{,}700 \times 10^6 \text{ Btu}$$

$$\text{savings in dollars} = (16{,}700 \times 10^6 \text{ Btu})(\$4.50/10^6 \text{ Btu})$$
$$= \$75{,}150/\text{year}$$

The savings for thermostat setback can be substantial, as shown in the example. In warm climates similar savings can be made by turning off the air conditioners at night and on weekends, but the dollar amount is usually less since heating demands peak at night, while cooling demands peak during the day.

Of course, the energy manager needs to be careful to ensure that the night setback doesn't

1. Effect process problems, e.g., tolerances on large metal parts or thermocure resins.

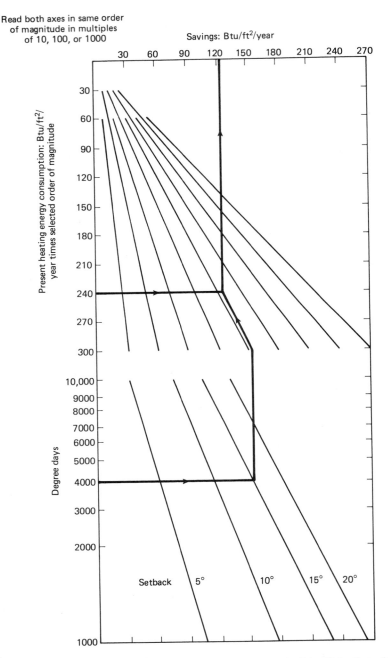

Figure 9–1 Estimation of savings via thermostat night setback. (*Identifying Retrofit Projects for buildings,* FEA/D-76-467, Sept. 1976.)

2. Effect heating plant problems. Night setback can sometimes cause air conditioning equipment to turn on. This depends on the particular equipment and controls used as well as the thermal load of the building itself.

Finally, night setback can often be applied to process areas. For example, large furnaces such as brick kilns should be turned off when possible, but often the preheat time required and/or the thermal wear on the furnace walls makes this impossible. Many times, however, the thermostat can be adjusted downward significantly without causing these problems. Often, trial and error is required to determine the optimum setting.

9.1.2 Timers

The next step in level of control complexity is the use of timers. These timers can range from very simple clocks to fairly complicated central time clocks with multiple channels for controlling numerous pieces of equipment on different time schedules.

For example, thermostat setback can be done manually as shown in Section 9.1.1 or automatically. Automatic controls can range from simple thermostats each with a built-in time clock (costing somewhere around $100 each), or a central time clock that overrides all the thermostats can be used. An installed single-channel central time clock will cost around $1000, but it can control numerous thermostats if all are on the same schedule. Different setback schedules require multiple channels, increasing the cost somewhat.

Some companies have utilized time clocks to duty cycle equipment such as exhaust fans. For example, a large open manufacturing area will likely have several exhaust fans. If there are six fans, then a central time clock could turn one fan off each 10 min and rotate so that each fan is off 10 min of each hour but no more than one fan is ever off at the same time. This saves on electrical consumption (kwh) to run the fan, electrical demand (since one fan is off at any time), and heating (since less conditioned air is exhausted). General ventilation over a wide area is maintained. Of course, care must be taken to ensure that no ventilation problems develop. See Problem 9.1 at the end of this chapter for an example calculation.

The use of timers allows a company to start-stop equipment at exactly the correct time. It is not necessary to wait for maintenance people to make their rounds, turning off equipment and adjusting thermostats. Timers don't forget, either.

On the other hand, timers do offer other problems. For example, power outages may require timers to be reset unless a battery backup is used. Also, daylight saving time changes require all timers to be set up an hour in the spring and back an hour in the fall. Finally, the clocks must be maintained and replaced as they wear out.

The authors had the opportunity to audit a plant that had sophisticated time clock controls on its equipment, but management was not maintaining the clocks. The 7-day time clocks allowed for night and weekend setbacks. The audit was done on a Thursday, but the time clocks read Saturday. Consequently, the thermostats were on night setback, and the people were cold. To remedy this, maintenance had purchased several additional portable heaters. If they had come in on a Saturday, when the clock read Monday, the plant would have been nice and warm. In this case, the poorly maintained clocks were costing the company a great deal of money. *Timers and any type of control system must be maintained.*

A type of control closely related to timers that has been receiving considerable attention recently is dimmers. Dimmers can be manually controlled or automatically controlled depending on time and natural lighting. It is important to be sure the dimming system chosen actually reduces electrical consumption and is not simply a rheostat that consumes the same amount of energy regardless of the amount of light delivered.

Supermarkets can often use relatively sophisticated dimming systems. For example, supermarkets might

1. Use photocells to detect natural light and dim window lights as appropriate.
2. Use photocells that determine dawn and dusk to turn parking and security lights on and off.
3. Use photocells to determine dusk so that interior lighting can be reduced. (Studies have shown that people coming from a dark street to a brightly lit room are actually uncomfortable. Lower lighting levels are preferred.)
4. Use demand sensors (to be covered) to turn off display lighting during peak demand times.

As with timers, photocells must be maintained. They sometimes fail, and if undetected, failure can cause other more severe problems. A regular maintenance schedule of checking photocells should be used.

9.1.3 Programmable Controllers

A programmable controller is a control device that has logic potential but is not powerful enough to be called a computer or even a microprocessor. As might be expected, it fills a need for systems requiring more than a timer but not a computer. It can do all that timers can do and considerably more but at a cost significantly less than that of a computer. One article [1] suggests that fixed logic devices (such as timers) are useful in buildings up to about 100,000 ft^2 and 300 control points. Computerized systems would be applicable in buildings of 200,000 ft^2 and more with 1000 control points. The article says that the rest would be suitable for programmable controllers.

A programmable controller adds logic capability to control systems. Demand shedding is a prime example. When the controller senses that the electrical demand is approaching a critical (programmed) level, the unit then shuts off equipment and/or lights to keep the demand from passing that critical level. As shown in Chapter 3, demand can be a big part of an electrical bill, so savings can be significant. (See Section 9.1.4 for more discussion on demand control.)

Another example is excess air control for a boiler or any larger combustion unit. By sensing CO_2 or O_2 and perhaps CO levels in exhaust, the controller can adjust the combustion air intake to yield optimum combustion efficiency. As shown in Chapter 7, this can be a real money saver. Continuous control through the use of a programmable controller allows the air intake to be adjustable according to demand on the unit.

Still another example where programmable controllers can be used is outside air control on heating, ventilating, and air conditioning systems. In air conditioning, outside air may be considerably more comfortable than inside air, so outside air should be used rather than returning inside air. In fact, sometimes the air conditioning units can be turned off completely and outside air used for cooling. Programmable controllers can sense outside vs. inside air enthalpy and determine the optimum damper setting. The same controller can shut off outside air completely for early morning start-up and nighttime operation during heating seasons. (See Section 9.1.4 for more on outside air control.)

In summary, programmable controllers are useful when logic is required, eliminating the use of timers, but the requirements are not large or complex enough to justify a computer.

9.1.4 Computerized Systems

Most computerized energy management control systems (EMCSs) sold today are microprocessor based. Capability runs from a few control points up to several thousand, with the larger ones often performing fire-safety functions and report generation as well as energy management. The technology is changing very rapidly as there are many vendors in the field each introducing new equipment. A potential user should consult several vendors and be well prepared to discuss needs.

Some additional control options are available to the EMCS user, but mostly they are the sum of all the techniques discussed in the previous sections. Figure 9-2 is useful in summarizing these techniques.

In Figure 9-2(a), the original electrical demand profile is shown (before control is applied). Note that the area under these curves is the integration of demand over time and thus is the kwh consumption.

In Figure 9-2(b), demand control is applied. Here, a peak demand is determined, and loads are shed once that peak is approached. Shedding requires predetermining what loads can be shed and in what priority. For example, display lighting would likely be shed before office lighting. Many systems also "remember" loads previously shed, and will rotate among shedable loads, and will obey preset maximum shed times. As Figure 9-2(b) shows, shed loads must sometimes be recovered and sometimes not. For example, shedding refrigeration saves demand, but sooner or later the unit must catch up. Shedding lights, on the other hand, cannot be recovered, so energy is saved. As shown, then, some shifted consumption occurs but usually not as much as was shed. Demand savings remain the predominant goal.

In Figure 9-2(c), fixed start-stop is utilized. Now units are turned on and off at exactly the same time and when needed each day. No longer are personnel required to make rounds, turning equipment on and off.

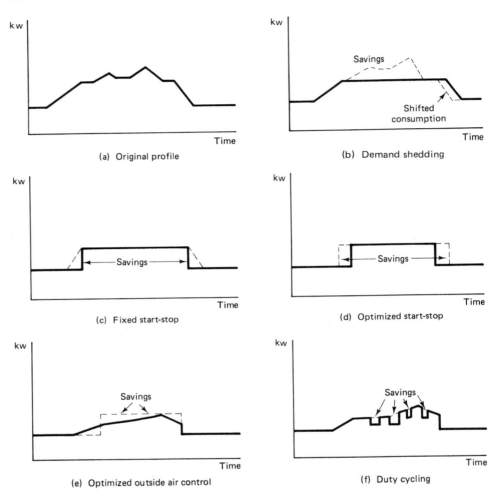

Figure 9-2 EMCS control techniques. (From Dick Foley, ''Reducing Waste Energy with Load Controls,'' *Industrial Engineering,* July 1979, p. 24. Extracted with permission from Institute of Industrial Engineers, Inc., 25 Technology Park/Atlanta, Norcross, Ga., 30092, © 1979.)

In Figure 9-2(d), optimized start-stop is employed. The precise time of need is determined each day, and the equipment is turned on at that time. For example, if the outside and inside temperatures are warm, the heating units do not have to be turned on as early as they would be if the respective temperatures were quite cold.

Figure 9-2(e) shows what happens when the use of outside air is optimally controlled. In this case, the plant requires air conditioning. The fans are turned on early (but maybe not the compressors) to draw in cool outside air for precooling. As the daytime temperatures warm, less outside air is used, and the compressors have to run longer until at the peak air conditioning time a minimum of outside air is used. Toward the end of the operating time, it may again become profitable to utilize outside air.

Duty cycling is depicted in Figure 9-2(f). Here loads are selected to be turned off on a predetermined schedule. For example, exhaust fans (see earlier discussion) may be turned off 10 min out of each hour. If these schedules are determined so something is being cycled at any time when peaks are likely to occur, demand savings will also develop. Figure 9-2(f) does not show any demand savings.

All the techniques shown in Figure 9-2 are easy to show in a figure, and all deal with electrical consumption. Other techniques cannot be easily demonstrated in such a figure because of their nature and/or because they affect fuel consumption rather than electricity. Examples include the following:

1. Light dimming, as discussed in Section 9.1.2.
2. Combustion air control for furnaces. This affects fuel consumption.
3. Night setback for heating normally affects fuel consumption instead of electricity.
4. Surge protection. If power outages occur, EMCSs can be programmed to start turning loads off to prevent an extremely large surge of power once the service is reconnected. This is impossible to show in Figure 9-2.
5. Temperature reset. Here the temperature of supply air or water is modified to meet actual demand. For a heating system, the supply air temperature may be reduced by 10 to 20°F when heating demands are small. This could save substantial fuel.

It's important that EMCS users recognize the need for feedback. The computer may have sent a signal to turn off a load, but was it turned off? Some sensor is required to feed back the control status to the computer. Also, it is helpful if the computer maintains a record of control exercised. Then, histograms can be developed periodically to show how frequently any given load is being shed.

What types of computer systems are available? Below we will attempt to aid the reader in understanding the market as it exists at the time of this writing. The reader should remember that this market is changing very rapidly and that "staying up to date" is vitally necessary.

The first consideration (after defining the controls desired) in the design of an actual EMCS is what configuration to use. In a centrally controlled system, control is vested in one central unit—a microprocessor or minicomputer. The control points are accessed directly (star network) or through common wiring (common data bus network). These are shown in Figure 9-3. In the star network, control is more direct but installation considerably more expensive. Obviously, its utilization is limited to smaller facilities. The common bus design allows for common use of wiring, so its installation cost is less for large facilities.

Instead of a centrally controlled EMCS, many companies have gone to

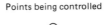

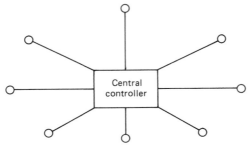

Star network

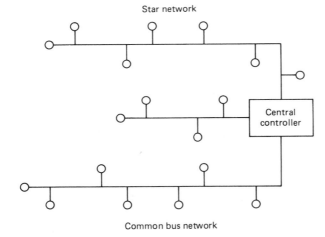

Common bus network

Figure 9-3 Centrally controlled EMCS
star network and common bus network.

distributed controls where a central unit is used primarily for coordination and report generation. This is demonstrated in Figure 9-4 using a star network.

The remote control units can have a varying amount of capability from a simple transfer function to complete control. In a true distributed control system, the remote controllers can function with or without the central unit—at least for a period of time. Of course, intelligent remote controllers are more expensive than unintelligent ones, but this appears to be changing. Most designers predict that future EMCSs will have more distributed control as the cost of remote controllers is reduced.

Of course, systems can be hybrids in that some star networking is used along with common bus designs. Also, some remote controllers may be intelligent and others not, and some points may be directly controlled by the central controller even in a distributed control basic design.

Another important factor to consider is the type of control needed at the points being controlled. The simplest and cheapest is a digital point which is a simple on-off control. Examples include switches for fans, lights, and motors. Analog controls, on the other hand, are more complex and therefore more expensive. Analog controls are

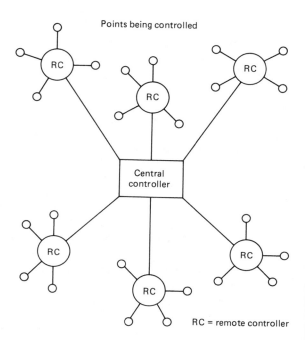

Points being controlled

Central controller

RC = remote controller

Figure 9-4   Distributed control EMCS
star network.

needed when signals of varying intensity are required. Outside air control for air conditioning systems is a good example. As the outside air cools, the EMCS will continue
to open the damper, allowing more outside air to enter. A significant number of
analog points will run up the cost of an EMCS rather rapidly, but the savings should
also increase. Careful studies should be made.

Some EMCSs fail for various reasons, but the reasons can usually be summarized into one of the categories shown in Table 9-1. The prudent energy manager will
consider these potential pitfalls and plan accordingly.

TABLE 9.1   WHY EMCS UNITS FAIL

1. *Inadequate buyer commitment.* All too often, buyers expect EMCS units to install, program, and maintain themselves. This simply will not happen.
2. *Inadequate seller backing.* The energy manager should ensure that the seller will back the product and
   provide the necessary technical aid.
3. *Poor vendor assessment.* The energy manager should screen the vendors carefully. Ask for reference
   letters and check with other energy managers. Be wary.
4. *Simple things not done first.* The EMCS can do many things, but it should not be asked to do the things
   that should have already been done. For example, energy-efficient lights cannot be installed by the
   EMCS.
5. *Simple alternatives overlooked.* An EMCS may not be necessary. Manual or timer control may suffice.
6. *Requirements not defined.* This is the most important reason. The buyer must define the requirements
   before choosing the system. Considerable thought is necessary.

Source: Drawn in part from "Energy Management Systems: A Critical Analysis" [2].

In summary, EMCSs can do many things, but they are only machines. The energy manager must be aware of their limitations as well as strengths and design accordingly. The major part of the design should occur before selection of the equipment—not afterwards.

9.2 COMPUTER UTILIZATION

Computers are used for everything today, so the energy manager must be aware of computing abilities and work computers into his or her job environment. Applications run all the way from small desk-top microprocessors programmed to run cash flow analyses, waste heat recovery studies, excess air control studies, and a myriad of other aids to large control computing systems capable of running large programs requiring significant data storage and computing time.

It is impossible to summarize or list all the possible uses, but one does stand out from the rest. That is the use of computers to run energy simulation studies. In the rest of this chapter we will examine some of the programs available.

Energy simulation studies involve definition of weather data, operating times, and other energy-consuming parameters such as number and types of lights and equipment, efficiency of various devices, etc. The computer will then *simulate* a year (or whatever cycle is chosen), developing energy consumption and energy bills. Thus, various scenarios can be fed into the computer and the likely savings identified or estimated. Most of the programs available also contain a financial analysis subroutine, so the complete study can be done by the computer. Some of the better known programs include BLAST 3.0, DOE 2.1A, TRACE, and E Cube, although there are many others. A call to your local utility, university, and/or consultant can identify which are available in your area. We will examine the capabilities of BLAST 3.0 and DOE 2.1A.

9.2.1 BLAST 3.0

BLAST can be used to investigate the energy performance of new and retrofit building design options of almost any type and size. Not only can BLAST calculate the peak load using design day criteria, which is necessary for mechanical equipment design; it can also estimate the annual energy performance of a facility. This is essential for the design of solar and cogeneration systems and for determining the compliance with a design energy budget.

Apart from its comprehensiveness, the BLAST system differs in three key aspects from similar programs. First, BLAST uses extremely rigorous and detailed algorithms to compute loads, to simulate fan systems, and to simulate boiler and chiller plants. Second, the program has its own user-oriented input language and is accompanied by a library which contains the properties of all materials, wall, roof, and floor sections. Third, BLAST's execution time is short enough to allow many alternatives to be studied economically. In this way, efficient designs can be separated from the inefficient; proper equipment type, size, and control can be determined.

The BLAST Energy Analysis Program contains three major subprograms. First, the Space Load Predicting subprogram computes hourly space loads in a building based on weather data and user inputs detailing the building construction and operation. Next, the Air Distribution System Simulation subprogram uses the computer space loads, weather data, and user inputs describing the building air-handling system to calculate hot water, steam, gas, chilled water, and electric demands of the building and air-handling system. Finally, the Central Plant Simulation subprogram uses weather data, results of the air distribution system simulation, and user input describing the central plant to simulate boilers, chillers, on-site power-generating equipment, and solar energy systems and computes monthly and annual fuel and electrical power consumption.

Users may obtain access to the BLAST program from various commercial vendors. Some of these vendors also offer user support and training for the system. Some of the current BLAST vendors include Boeing Computer Services, Control Data Corporation CYBERNET, and McDonnel Douglas Automation.

The BLAST system of programs is written in FORTRAN IV with many of the features specific to Control Data Corporation (CDC) computers. Therefore, BLAST is currently available only for CDC computers. However, a version adaptable to the IBM computer is currently in progress and will probably be available by the time this book is in print.

Additional information regarding BLAST 3.0 is available through the National Technical Information Service (NTIS), 5285 Port Royal Road, Springfield, VA 22161, and the Department of the Army, Construction Engineering Research Laboratory, P.O. Box 4005, Champaign, IL 61820.

9.2.2 DOE 2.1A

The DOE 2.1A Computer Simulation Program is used to simulate the energy use of plants and buildings. The program is divided into five major subprograms: (1) Building Description Language (BDL), (2) LOADS, (3) SYSTEMS, (4) PLANT, and (5) ECONOMICS.

The Building Description Language (BDL) subprogram allows the user to enter key building design information. The program uses a library of properties of all materials, walls, roof, and floor sections. The user also inputs a description of the HVAC systems, occupancy, equipment, lighting schedules, and other parameters.

The LOADS subprogram computes hourly space loads resulting from transmission gains and losses through walls, roofs, floors, doors, and windows; internal gains from occupants, lighting, and equipment; and infiltration gains and losses caused by pressure differences across openings. The LOADS calculations are based on ASHRAE algorithms, including the response factor technique for calculating transient heat flow through walls and roofs and the weighting factor techniques for calculating heating and cooling loads.

After the building loads are calculated, the program begins the SYSTEMS analysis. The SYSTEMS subprogram takes the hourly space loads, along with characteristics of secondary HVAC equipment, the component and control features, and the thermal characteristics of the zone, and determines the actual room temperature and heat extraction or addition rates using ASHRAE algorithms.

The PLANT program uses the building thermal energy load data determined by SYSTEMS and other various, user-inputed, operating parameters of the plant equipment to allocate available equipment and simulate their operation. The PLANT program simulates conventional central plants, solar heating and cooling systems, and plants with on-site generation and waste heat recovery. It also permits load management of plant equipment and energy storage. It calculates the monthly and annual cost and consumption of each type of fuel used, the daily electrical load profile, and the energy consumption at the site and at the source.

The ECONOMICS subprogram uses the life-cycle costing methodology derived from DOE guidelines. Life-cycle costing method investment statistics (i.e., cost savings, savings-to-investment ratio, energy savings, energy savings-to-investment ratio, and discounted payback period) are calculated to provide a measure for comparing the cost effectiveness of each case against a reference case.

Users may obtain access to the DOE 2 program from various commercial vendors. Some of these vendors also offer user support and training for the system. The DOE 2.1A program is currently available for the CDC 7600, CYBER 175, and IBM 370 and 303X. The program uses FORTRAN (99.9%) and COMPASS (.1%) for the CDC version. The IBM version is all FORTRAN.

Additional information on DOE 2.1A is available through the National Technical Information Service (NTIS), 5285 Royal Road, Springfield, VA 22161, and the Building Energy Analysis Group, Energy and Environment Division, Lawrence Berkeley Laboratory, Berkeley, CA 94720. Information is also available through the Department of Energy, Office of Conservation and Solar Applications, Division of Buildings and Community Systems, Architectural and Energy Systems Branch.

9.3 SUMMARY

In this chapter we have examined control systems and computer applications for energy management. We began with a discussion of the range of controls, including manual, timer, programmable controllers, and computers. Then we discussed each level of control, giving advantages and limitations. Bascially, the simpler controls are less expensive and less robust. The more expensive controls (such as EMCSs) are more robust in that more control activities can be utilized.

Manually controlled systems can turn equipment on and off, set back thermostats, and in general control equipment. Reliability and timing are often problems. Timers ensure proper timing and improve reliability but add considerable maintenance. Photostatic dimming is another type of timing control that has proven useful in many buildings.

Programmable controllers are essentially small microprocessors that can do all the preceding plus add some limited logic capability such as excess air control, demand control, and outside air control. Computerized systems are the top of the line in that all the control schemes mentioned so far can be utilized with *practically* no limitation on size. In addition, computerized systems can aid in monitoring, cost center metering, and report generation. Obviously, the first cost for the larger computerized systems is quite high.

Computers can be used in other areas of energy management also. Data manipulation, data summary, and large-scale modeling or simulation are only examples of other areas where computers can be utilized. Large-scale computer simulation models of energy systems are available and are quite useful in simulation, various scenarios of new equipment, or revised control schemes. BLAST 3.0 and DOE 2.1A are two programs discussed in some depth in the chapter.

REFERENCES

1. "Programmable Controllers: A Viable Alternative," *Plant Energy Management,* March 1982, pp. 44–45.
2. Gary G. Bullock, "Energy Management Systems: A Critical Analysis," *Energy Engineering,* Aug.–Sept. 1982, pp. 21–36.

BIBLIOGRAPHY

BLAKE, FREDERICK H., "New Energy Management Controls for Supermarkets," *Energy Engineering,* Aug.–Sept. 1981 pp. 49–54.

"Building Loads Analysis and Systems Dynamics (BLAST)," Fact Sheet, U.S. Army Corps of Engineers, Construction Engineering Research Laboratory, Champaign, Ill., 1978.

FOLEY, DICK, "Reducing Waste Energy with Load Controls," *Industrial Engineering,* July 1979, pp. 23–27.

TURNER, WAYNE C., Senior ed., *Energy Management Handbook,* Wiley-Interscience, New York, 1982.

QUESTIONS

9.1. In the section on demand control, the discussion said that some loads must be recovered (i.e., run later) and some not. Give an example of a load that must be recovered, one that doesn't need recovery at all, and one that may need partial recovery.

9.2. Discuss uses of computers in energy management other than direct control.

9.3. You have just finished auditing a large supermarket that operates 16 h/day. The market has substantial glass exposure to the outside and substantial lighting for display purposes. Outside lights are used for parking and security. Forgetting any change of light sources, what control schemes would you recommend?

9.4. Someone once said that improperly maintained timers can cost more energy than they save. Discuss.

9.5. Discuss examples of loads whose start-stop timers can be optimized as in Figure 9-2(d).

PROBLEMS

9.1. Ugly Duckling Manufacturing Company has a series of 12 exhaust fans over its diagnostic laboratories. Presently, the fans run 24 h/day, exhausting 600 cfm each. The fans are run by $\frac{1}{2}$-hp motors. Assuming the plant operates 24 h/day, 365 days/year in an area of 5000°F heating degree days and 2000°F cooling degree days, how much will be saved by duty-cycling the fans such that each is off 10 min/h on a rotating basis? At any time, 2 fans are off and 10 are running. The plant pays $.05/kwh and $5.00/kw for its electricity and $5.00/10⁶ Btu for its gas. The heating plant efficiency is .8, and the cooling efficiency is 2.5.

9.2. Profits, Inc. has a present policy of leaving all its office lights on for the janitor crew at night. After a careful analysis, the company finds it can turn off 1000 fluorescent light bulbs (35 w each) at closing time (4 h earlier than before) and still leave enough on for the janitorial crew. Assuming the company works 5 days/week, 52 weeks/year, what is the saving for turning these bulbs off an extra 4 h/day? The company pays $.06/kw and $6.00/kw for electricity. Peaking hours for demand are 1:00–3:00 p.m.

9.3. Therms, Inc. has a large electric heat-treating furnace that takes considerable time to warm up. However, a careful analysis shows the furnace could be turned back from a normal temperature of 1800°F to 800°F 20 h/week and be heated back up in time for production. If the ambient temperature is 70°F, the composite R value of the walls and roof is 12, and the total surface area is 1000 ft², what is the savings in Btu for this setback?

9.4. Obtain bin data for your region, and calculate the savings in Btu for a nighttime setback of 15°F from 65 to 50°F 8 h/day (midnight to 8:00 a.m.).

9.5. Petro Treatments has its security lights on timers. The company figures that by using photocell controls an average of 1 h/day that operating time can be saved. The company has 100 mercury vapor bulbs of 1000 w each. If the company pays $.06/kwh, what are the savings? Assume there are no demand savings.

Chapter 10

· ·
Maintenance

10.0 INTRODUCTION

Maintenance is affected by, and affects, every major energy management measure. Maintenance personnel replace lamps, change settings on boilers, and fix leaking steam traps. If a system is well maintained, most of the easy energy management measures have already been done, and the maintenance crew is interested in finding other measures. The crew works well with line management, production stoppages for nonroutine maintenance measures are scheduled in advance, and few unexpected breakdowns can be attributed to poor maintenance. In this situation, energy management becomes another routine part of a good maintenance program.

In many facilities, however, maintenance is an afterthought, and the people who perform maintenance are untrained, receive low pay, and have a low status within the company. In such facilities, the maintenance personnel are often the most expensive employees because production breakdown and energy waste can be traced directly to the maintenance program (or lack thereof).

We have two objectives in this chapter: to show how to set up a good maintenance program for energy management systems and to show how to make a good program better. The first step in meeting these objectives is to assess current facility needs for repairs and for routine maintenance. Then a method is presented for determining how often routine maintenance should be performed and how long each maintenance action should take. The total demand for maintenance dictates the number and skill levels of people on the maintenance crew and their support equipment.

10.1 DETERMINING MAINTENANCE NEEDS

10.1.1 Repair Maintenance

In performing the energy audits described in Chapter 2, each of the major energy-using systems is examined in detail. As these systems are inspected, notes should be taken on the maintenance measures needed to bring each into satisfactory condition. In some cases, determining what is wrong is difficult, and estimating the time needed to fix it can be even more so. Fixing the problem can be complicated by a lack of parts and by an unwillingness on the part of maintenance people to admit they don't know how to repair a given unit. The authors saw a circulating pump where the pump motor had been removed for repair 2 years before we arrived and had not been either repaired or replaced in the intervening interval. Meanwhile, rooms on the periphery of the building had 45°F temperatures, and rooms on the interior had 95°F temperatures. Similar situations occur frequently, and one of the functions of the energy audit is to identify them.

10.1.2 Routine Maintenance

In addition to the repair needs shown by the energy audit, it is necessary to determine what preventive measures should be scheduled on a regular basis. In Chapter 5 we spoke about three causes of lighting quality deterioration—dirt, lamp dimming, and lamp wearout—and analogies to these can be found in most systems. Preventive maintenance should aim to (1) detect a deteriorating condition—as with a light meter, (2) perform cleaning tasks—as with lubrication or cleaning light fixtures, (3) replace worn parts—bearings or lights, and (4) assess the general condition of the system—noting that a motor is overheating or that a ballast is beginning to smoke. Preventive maintenance helps to lengthen the life of equipment so that major repairs are not necessary as often and helps to detect major problems early enough that repairs can be scheduled at a convenient time.

What preventive maintenance tasks should be done? Vendors are a good source for this information. So also are such books as that by Sack [1] and the Department of Defense facilities maintenance manual [2]. These books give a guide as to the time, skill level, and equipment needed for routine maintenance. In addition, equipment manuals should be kept and their instructions observed. Special procedures can be used for some items such as bearings; the Tornberg procedure for bearings [3] is an example. (For motors that are driving either fan belts or chains, use an industrial stethoscope to check both ends while the motor is in use. The driving end should be 2–3 db louder than the fixed end. If it is more than 2–3 db louder, something is wrong with the bearings in the driving end. If the difference is less than 2–3 db or the fixed end is louder, then bearings are going bad in the fixed end.) Other sources for maintenance information are professional societies, such as the Institute of Industrial Engineers or the American Institute of Plant Engineers, and trade journals. (*Power* magazine, published by McGraw-Hill for the utility industry, for example, has a regular section on maintenance.)

As maintenance procedures are developed, file cards should be prepared on every piece of equipment showing bearing numbers, equipment serial numbers, fan belt sizes and serial numbers, and other information unique to the particular item of equipment. These cards are invaluable when something breaks down or when ordering spare parts.

Maintenance needs differ from company to company, depending on the abuse received by the equipment, dust in the atmosphere, temperature swings, and past maintenance practices. As a consequence, it is desirable and perhaps necessary to adapt for each new environment the maintenance experience used elsewhere. The objective is the same: increased production through wise maintenance.

10.2 DETERMINING MAINTENANCE INTERVALS

The determination of maintenance intervals is governed by three principles. First, the more often an equipment item is maintained, the higher the direct cost of maintenance labor. Second, the more often an equipment item is inspected, the more likely it is that equipment defects will be detected in time to schedule repairs at a convenient time (unless the inspection itself contributes to the problem). Third, if preventive maintenance is perceived as a detriment to production, it will be postponed or simply will not be performed.

It would be convenient to have curves showing performance degradation over time and to schedule maintenance so that the performance never drops below a prescribed level. This can be done to a certain extent in lighting, as shown in Chapter 5. In most situations, however, the interval between inspections and regular preventive maintenance actions must be determined on the basis of experienced need. In a dusty environment, for example, it may be necessary to change air filters once each month. In a clean environment, it may be possible to get by with checking filters only once each year, replacing them only every 2 years. Some criteria must be established to define when a filter is dirty, and they depend on experience.

The choice of a maintenance interval is easier to defend if it is based on some kind of measurement. If, for example, a filter is dirty at the same time there is a drop of 1-in. water gauge pressure across it, then a manometer can be installed, the pressure drop can be recorded every week or month, and a preventive maintenance interval can be chosen so that the pressure drop never exceeds 1-in. water gauge. (See Problem 10.1.)

Maintenance intervals can also be based on energy consumption if the change of energy consumption with time can be determined. If, for example, a steam system is old and has not been maintained very well, then the number of steam leaks that appear between inspections may be proportional to the length of the inspection interval, and the steam loss will increase with the length of this interval. One area in which loss in energy efficiency is particularly noticeable is in boiler maintenance—controls that

Problem	Solution	Cost/benefit[a]
Building envelope (outside walls and roof):		
Doors: Loose fitting	Weatherstripping	VL/LA
Do not close	Correct fit; check pressures	L/LA
Windows: Air leakage	Weatherstrip, caulk	VL/LA
Broken	Replace panes	L/MA
Walls: Cracks	Patch or seal	M/H
Roofs: Holes	Patch or cover	M/M
Boiler and steam distribution system:		
Boiler: Safety problems	Correct *now*	S
Inoperable gauges	Overhaul	VH/VH
Stack gas clear or black	Check and adjust	H/VH
Scale deposits	Fix water treatment	H/H
Steam traps leak	Repair or replace	L/H
Steam leaks	Repair	H/VH
Steam line insulation ragged or missing	Repair or replace	L/M
Condensate return system uninsulated but hot	Insulate	L/M
Heating, ventilating, and air conditioning system:		
Filters very dirty	Replace or clean	L/LA
Dampers blocked or disconnected	Repair linkage and check controls	L/H
Ductwork blocked inside	Examine and repair	L/LA
Grillwork blocked or dirty	Remove blockage; clean	VL/MA
Fan motor not connected	Replace belts or disconnect motor	VL/M
Thermostat reading not accurate	Calibrate and correct	VL/LA
Electrical system:		
Transformers:		
Leak	Replace immediately	S
Heat transfer surfaces fouled	Clean carefully	L/L
Contacts show burned spots	Repair immediately	S
Frayed wire	Tape carefully	S
Switches arc	Replace	L/M
Lights, windows, and reflective surfaces:		
Unused space illuminated	Remove lights or install better switches	VL/L
Ballasts buzz	Adjust	L/LA
Lights flicker	Replace or eliminate	VL/LA
Walls dark	Use light colored paint; clean walls	VL/LA
Windows dirty	Clean or paint over	L/L
Hot water distribution system:		
Water too hot	Lower thermostat setting	VL/M
Faucets or lines leak	Repair or replace	L/M
Water heated when not needed (weekends)	Install or reset time clocks	L/H
Scale buildup	Check water treatment system	M/MA

Figure 10–1 Maintenance for the major energy systems. (W. J. Kennedy, Jr., "Apply a Cost-Benefit Approach in Building an Energy System Maintenance Program," *Industrial Engineering,* Dec. 1981, pp. 62–63. Reprinted with permission from the Institute of Industrial Engineers, Inc., 25 Technology Park/Atlanta, Norcross, Ga. 30092, © 1981.)

Problem	Solution	Cost/benefit[a]
Air compressors and compressed air distribution system:		
Compressor air leaks	Overhaul	M/L
Compressor oil leaks	Overhaul	M/H
Oil in pneumatic control system	Overhaul compressor	M/VH
Air line leaks	Repair	VL/M
Manufacturing system — processes:		
Controls not working	Overhaul	VH/VH
Working temperatures too high	Study and correct	H/VH
Motors noisy	Check bearings	L/H
Motors hot	Adjust voltages on legs of three-phase input; correct belt-tension	M/M
Oven refractory problems	Repair or replace	M/H
Oven gaskets leak	Replace	VL/M
Material handling systems:		
Drafts	Check envelope, pressure balance (inside-outside)	L/MA
Uncontrolled fuel use	Institute accounting of fuel per vehicle	L/MA
Excessive tire wear	Check tire pressure	L/M
Excessive fuel use	Keep vehicles tuned up	L/MA
MH system runs when nothing is being carried	Check load switches regularly	VL/H
Excess use of diesel fuel	Follow manufacturer's recommended procedures; if timed switches limit idling, check starters often	

[a] Key:

Costs: VL = less than $25
          L = $25–$200
          H = $200–$1000
         VH = more than $1000
         M = $100 = 300

Benefits: L = less than $25/year
         M = $25–$150
         H = more than $150
         A = substantial noncost benefits (comfort, appearance, etc.)
         S = safety-related

Figure 10-1   *(Cont.)*

are not inspected and maintained regularly get out of adjustment, with a resulting loss of efficiency, fuel-handling equiipment gets worn, and burners get out of adjustment. (A rule of thumb: If a boiler has not been maintained for 2 years, a 30% gain of efficiency is immediately possible with maintenance.) In most other systems, a lack of maintenance is also associated with energy consumption. In HVAC systems, controls drift out of calibration, motors wear and shift on their mounts, and all the air-conducting components get dirty. In hot water systems, washers get worn, and faucets start to leak. In industrial processes, the heat transfer surfaces on waste heat recovery units get fouled, control systems drift out of calibration, and oven gaskets get worn. Similar, longer lists can be made for each system. One example of such a list is shown in Figure 10-1 [4].

Another point that should also be taken into consideration in determining maintenance intervals is the likelihood of performance loss in the energy-consuming systems. A steam leak costs money in lost energy, and it also costs money in degraded performance in the process using the steam. In a plywood plant where steam is used to dry plywood veneer, for example, a loss of steam was directly associated with a loss of productive capacity even though the steam cost was very small (steam was generated by waste wood products). In the compressed air distribution system, leaks cause the compressor to work more and use more energy, but they also may cause problems with the control system using the compressed air. Thus performance degradation should be taken into account in choosing maintenance intervals.

Any system whose degradation can create safety problems should be inspected regularly regardless of other considerations. This is a primary responsibility of management and must be a paramount concern to any professional.

In many systems, performance degradation is gradual, and the optimum interval is some time in a large range, for example, 4–8 weeks. The choice of a maintenance interval can then be made for the convenience of the maintenance staff so that similar maintenance tasks can be performed at the same time and the same economies can be used that make group relamping attractive. If the maintenance is performed by company personnel, it is usually desirable to choose maintenance intervals so that the load on personnel is constant.

Once a maintenance schedule has been chosen, it should be communicated and checked. This can be done either through a computerized system with wide access to the communications data base or through a tickler file with one group of cards for each day. Either way, it is important to know what is to be done, what has been done, and whether the maintenance interval can be lengthened or should be shortened.

10.3 STAFFING AND SUPPORT

It is difficult to determine staffing needs for maintenance. Three questions must be answered: How many people are needed, what skill levels should be employed, and how much should be invested in support equipment?

10.3.1 Estimating the Number of People Needed

The usual method for determining the number of staff members required for a manufacturing operation is to (1) calculate the number of operations that need to be performed, (2) estimate the standard time (including factors for personal time, fatigue, machine-caused delay, and transportation) for each operation, (3) add overhead, and (4) use the previous steps to estimate the number of man-years required per year. Then staff members are selected accordingly. For routine maintenance, this is possible, and References 1 and 2 give good guidance. In many systems, however, the major problem is repair rather than preventive maintenance, and a different technique must be used.

In estimating times for repair maintenance, it is necessary to take into account the experience of people doing the repair, the availability of necessary spare parts, the availability of diagnostic test equipment, and the possible cost of having the work done by outside contractors. Generally, a time estimate should be made by the person who is responsible for the work, with an evaluation of this estimate based on the maintenance person's past performance.

In addition to obtaining time estimates for repair maintenance, it is important to obtain cost estimates for repair equipment, spare parts, and outside expertise. It is equally important to estimate the costs of consequences if repairs are not performed. (Again, safety should be considered first.)

Once benefits and costs have been tabulated for each repair, the repairs can be prioritized and worked into the schedule of regular maintenance personnel. If the repair load is very large, it may be necessary to hire additional people, depending on the need and potential benefits.

There is a great temptation to avoid preventive maintenance so as to have more time for repair. This is understandable but doesn't really make much sense. Preventive maintenance increases the short-term work load and decreases the long-term work load. If such time-consuming items as bearing replacement can be regularly scheduled, then the number of in-plant bearing repairs, and the incidence of production stoppages due to bearing failures, will decline, and the repair load will also decline.

10.3.2 Skill Requirements

It pays to get the best possible people on maintenance and to pay them enough to keep them. A journeyman will have enough experience to perform nonroutine repair as well as to do the usual lubrication, cleaning, and routine maintenance; an apprentice may have the enthusiasm but will probably not have the skill. If employees are given specific training in energy management, they can add these tasks as well and, in the authors' experience, will do so gladly.

The psychology of dealing with maintenance personnel is important. If the most visible results of an energy auditor's visit are criticism of maintenance procedures and an increase in work load, the next visit may not be greeted with enthusiasm. Many energy management measures, however, reduce the maintenance workload, and such reduction should be pointed out. Special attention should also be given to the suggestions given by maintenance people; often the suggestions have been made before but ignored on the philosophy that (1) the person making them is in maintenance and therefore not worth listening to and (2) the person is working in *our* plant and therefore cannot be very smart. When suggestions from maintenance personnel are given support by the energy management team with full credit to their source, the team gains credibility (if the suggestions are good), and cooperation is measurably helped.

10.3.3 Support Equipment

For effective energy management maintenance, attention must be paid to support equipment, specifically including manuals, diagnostic equipment, spare parts inventories, and instruments for routine inspections.

10.3.3.1 Manuals.
Manuals are available from most vendors for most equipment items. One copy of each manual should be kept in a central location with an index system designed to facilitate its use. Other copies should be made available where they are needed. These manuals are a good source of basic maintenance material, including schedules and equipment.

10.3.3.2 Diagnostic equipment.
Good diagnostic equipment is a great aid in identifying the exact cause of a problem. What constitutes "good," however, depends on the quality of the equipment, the nature of the documentation, and the skill of the user. Sources of information on diagnostic equipment are vendors, training seminars, and basic experience of maintenance personnel. All these should be consulted, both formally and informally.

10.3.3.3 Spare parts inventories.
Enough spare parts inventories must be kept on hand to assure that maintenance can be performed either when planned or when immediate repairs are needed. But some parts are too expensive to stock routinely, and other parts do not break down, so there seems to be no point in stocking them at all. A thorough job of maintenance management includes keeping and analyzing

Instrument	Cost	Function
Digital thermometer with surface and probe attachments, 50–2000°F	$200–350	Check stack, process, and room temperatures
Vane anemometer	300–400	Determine velocity of air in ducts and through envelope gaps
Pitot tube anemometer	200–300	Measure velocity of flue gas or fluids in ducts
Industrial light meter	40–60	Measure light conditions
Combustion gas analyzer	700–900	Analyze composition of flue gas
Industrial stethoscope	300–400	Check bearings on motors
Clamp-on ammeter with chart recorder	600–800	Measure current over time; determine electrical peak demand times
Clamp-on volt-ohm-milliammeter	80–200	Measure voltage, current where needed; check for short circuits

Figure 10-2   Instruments for maintenance in energy management.

records on all equipment breakdowns and on needs for spare parts. It is also crucial to know who vendors are and what the lead times are for materials, both under routine and under emergency conditions. The objectives of these analyses are to ensure that all parts are on hand when maintenance is scheduled, to keep inventories of maintenance parts to a small dollar value but to have enough parts on hand for critical breakdowns, and to keep a ready supply of cooperative vendors available so that emergency parts can be obtained when needed.

10.3.3.4 Instruments for routine inspections. The authors consider it essential to have on hand a number of instruments for routine inspection of energy management systems. These instruments, their uses, and their costs are shown in Figure 10–2.

10.4 SUMMARY

In this chapter we have given a brief review of maintenance needed as a part of energy management. The development of a maintenance program should include some elements aimed at reducing energy costs and/or increasing the amount of production available from a given amount of energy. Serious thought should be given to the people involved in maintenance: how many, who, and how they should be treated. Sufficient support equipment must be provided as part of regular maintenance costs.

REFERENCES

1. Thomas F. Sack, *A Complete Guide to Building and Plant Maintenance,* Prentice-Hall, Englewood Cliffs, N.J., 1971.
2. *Unit Price Standards Handbook,* NAVFAC P–716 (or ARMY TB 420–33 or AIR FORCE 85–56), Nov. 1977.
3. Wayne C. Turner, Senior ed., *Energy Management Handbook,* Wiley-Interscience, New York, 1982.
4. W. J. Kennedy, Jr., "Apply a Cost-Benefit Approach in Building an Energy System Maintenance Program," *Industrial Engineering,* Dec. 1981, pp. 60–65.

QUESTIONS

10.1. What routine preventive maintenance tasks should be performed in a house gas furnace, and why aren't they?
10.2. What criteria should be used in determining priorities of repair maintenance projects? How would you weight these criteria?

10.3. With two other people, walk through a church or some building with systems in need of repair, and list specific repair jobs. Then make a list of criteria to be used in weighting these jobs, and weigh each job against each criterion. Then multiply the criteria weights by the job weights to get a weight for each job. Does the resulting ranking make sense? If not, find some way to improve this system.

PROBLEMS

10.1. In determining how often to change filters, an inclined tube manometer is installed across a filter. Conditions have been observed as follows:

Week	Manometer reading	Filter condition
1	.4 in. water	Clean
2	.6	Clean
3	.7	A bit dirty
4	.8	A bit dirty
5	.8	A bit dirty
6–9	.9	Dirty
10–13	1.0	Dirty
14–18	1.1	Dirty
19–23	1.2	Very dirty
24	1.3	Plugged up; changed

Based on this table, give a range of times for possible intervals for changing filters.

10.2. You have been keeping careful records on the amount of time taken to clean air filters in a large HVAC system. The time taken to clean 20 filter banks was an average of 20 min/filter bank and was calculated over several days with three different people—one fast, one slow, and one average. Additional time that must be taken into account includes personal time of 20 min every 4 h. Setup time was not included. Calculate the standard time for filter cleaning, assuming that fatigue and miscellaneous delay have been included in the observed times.

10.3. Your company has suffered from high employee turnover and production losses, both attributable to poor maintenance (the work area was uncomfortable, and machines also broke down). Eight people left last year, six of them probably because of employee comfort. You estimate training costs as $10,000/person. In addition, you had one 3-week problem that probably would have been a 1-week problem if it had been caught in time. Each week cost approximately $10,000. All these might have been prevented if you had had a good maintenance staff. Assuming that each maintenance person costs $25,000 plus $15,000 in overhead per year, how many people could you have hired for the money you lost?

10.4. A recent analysis of your boiler showed that you have 15% excess combustion air. Discussion with the local gas company has revealed that you could use 5% combustion air if your controls were maintained better. This represents a calculated efficiency improvement of 2.3%. How large an annual gas bill is needed before adding a maintenance person for the boiler alone is justified if this person would cost $40,000/year?

10.5. Your steam distribution system is old and has many leaks. Presently, your coal bill is $600,000/year. A careful energy audit estimated that you were losing 15% of the generated steam through leaks and that this could be reduced to 2%. What annual amount would this be worth, considering energy costs only?

Chapter 11

. . . .

Insulation

11.0 INTRODUCTION

The proper use of thermal insulation is vitally important to the successful operation of any energy management program. Good engineering design of insulation systems will reduce undesirable heat loss (gain) at least 90% in most applications and improve environmental conditions in many. Consequently, it is necessary to understand insulation theory and applications.

In the first part of this chapter we discuss the theory behind the calculations. There are two basic areas of insulation applications: buildings and process equipment. In the rest of the chapter we discuss insulation types available for buildings and for process applications.

Solar energy plays a vital role in energy gain control for both buildings and process equipment. Unwanted gains are usually controlled by overhangs or other blocking systems, films, and sometimes landscaping, which may be considered insulation. This area, however, will be covered in Chapter 13 on alternative energy sources.

11.1 THEORY: HEAT TRANSFER, THERMAL CONDUCTIVITY, RESISTANCE, AND TREATMENT OF PIPES

There are three basic modes of heat transfer. They are conduction, convection, and radiation. Each is defined as follows:

- *Conduction.* Heat transfer from a hot side to a cooler side through a dividing medium. The hot side heats the molecules in the dividing medium and causes them to move rapidly, heating adjacent molecules until the cool side is heated. Transfer stops when the temperature of the hot side equals that of the cool side.

- *Convection.* Heat transfer between a moving liquid or gas and some conducting surface. Usually the heated fluid rises, causing cooler fluid to come in contact, which is then heated and rises, etc.
- *Radiation.* Heat transfer based on the properties of light so no transfer medium is necessary. For example, the sun heats by radiation.

The thermal conductivity (K) of a material is its ability to conduct heat. It is measured by the amount of energy (Btus) per hour that can pass through 1 f^2 of surface 1 in. thick (be careful; sometimes tables show K values for 1-ft thicknesses) for a 1°F temperature difference between the two environments being separated. The units are as shown in Table 11-1, which presents K values for various materials at room temperatures. Thermal conductivity will vary with temperature, which can be important for process applications, as will be seen. For K values of common insulating materials at varying temperatures, see Figure 11-1.

The rate of heat transfer is directly proportional to the temperature difference as shown in Equation 11-1 and inversely proportional to the resistance against heat flow:

$$Q \propto \frac{\Delta t}{R} \qquad\qquad (11\text{-}1)$$

TABLE 11-1 THERMAL CONDUCTIVITY VALUES FOR VARIOUS MATERIALS AT ROOM TEMPERATURE

Description	$K \dfrac{(\text{Btu} \cdot \text{in.}}{\text{ft}^2 \cdot \text{h} \cdot °\text{F})}$
Aluminum (alloy 1100)	1536
Aluminum bronze (76% Cu, 22% Zn, 2% Al)	696
Brass:	
Red (85% Cu, 15% Zn)	1044
Yellow (65% Cu, 35% Zn)	828
Bronze	204
Copper (electrolytic)	2724
Gold	2064
Iron:	
Cast	331
Wrought	418.8
Nickel	412.8
Platinum	478.8
Silver	2940
Steel (mild)	314.4
Zinc:	
Cast	780
Hot-rolled	744

Source: Albert Thumann, *The Plant Engineer's and Manager's Guide to Energy Conservation,* © 1977. Reprinted with permission of Van Nostrand Reinhold Co., New York.

TABLE 11-1 (CONT.)

MATERIAL	DESCRIPTION	CONDUC-TIVITY κ#	CONDUCT-ANCE c + •
BUILDING BOARDS	ASBESTOS-CEMENT BOARD	4.0	
	GYPSUM OR PLASTER BOARD...1/2 IN.		2.25
	PLYWOOD	0.80	
	PLYWOOD...3/4 IN.		1.07
	SHEATHING (IMPREGNATED OR COATED)	0.38	
	SHEATHING (IMPREGNATED OR COATED) 25/32 IN.		0.49
	WOOD FIBER—HARDBOARD TYPE	1.40	
INSULATING MATERIALS	BLANKET AND BATT:		
	MINERAL WOOL FIBERS (ROCK, SLAG, OR GLASS)	0.27	
	WOOD FIBER	0.25	
	BOARDS AND SLABS:		
	CELLULAR GLASS	0.39	
	CORKBOARD	0.27	
	GLASS FIBER	0.25	
	INSULATING ROOF DECK...2 IN.		0.18
MASONRY MATERIALS	LOOSE FILL:		
	MINERAL WOOL (GLASS, SLAG, OR ROCK)	0.27	
	VERMICULITE (EXPANDED)	0.46	
	CONCRETE:		
	CEMENT MORTAR	5.0	
	LIGHTWEIGHT AGGREGATES, EXPANDED SHALE, CLAY, SLATE, SLAGS; CINDER; PUMICE; PERLITE; VERMICULITE	1.7	
	SAND AND GRAVEL OR STONE AGGREGATE	12.0	
	STUCCO	5.0	
	BRICK, TILE, BLOCK, AND STONE:		
	BRICK, COMMON	5.0	
	BRICK, FACE	9.0	
	TILE, HOLLOW CLAY, 1 CELL DEEP, 4 IN.		0.90
	TILE, HOLLOW CLAY, 2 CELLS, 8 IN.		0.54
	BLOCK, CONCRETE, 3 OVAL CORE:		
	SAND & GRAVEL AGGREGATE...4 IN.		1.40
	SAND & GRAVEL AGGREGATE...8 IN.		0.90
	CINDER AGGREGATE...4 IN.		0.90
	CINDER AGGREGATE...8 IN.		0.58
	STONE, LIME OR SAND	12.50	
PLASTERING MATERIALS	CEMENT PLASTER, SAND AGGREGATE	5.0	
	GYPSUM PLASTER:		
	LIGHTWEIGHT AGGREGATE...1/2 IN.		3.12
	LT. WT. AGG. ON METAL LATH...3/4 IN.		2.13
	PERLITE AGGREGATE	1.5	
	SAND AGGREGATE	5.6	
	SAND AGGREGATE ON METAL LATH 3/4 IN.		7.70
	VERMICULITE AGGREGATE	1.7	
ROOFING	ASPHALT ROLL ROOFING		6.50
	BUILT-UP ROOFING...3/8 IN.		3.00
SIDING MATERIALS	ASBESTOS-CEMENT, 1/4 IN. LAPPED		4.76
	ASPHALT INSULATING (1/2 IN. BOARD)		0.69
	WOOD, BEVEL, 1/2 X 8, LAPPED		1.23
WOODS	MAPLE, OAK, AND SIMILAR HARDWOODS	1.10	
	FIR, PINE, AND SIMILAR SOFTWOODS	0.80	
	FIR, PINE & SIM. SOFTWOODS 25/32 IN.		1.02

*Same as U value.
#Conductivity given in Btu•in./h•ft^2•°F
+Conductance given in Btu/h•ft^2•°F
Source: Extracted with permission from *ASHRAE Guide and Data Book*, 1965. Reprinted with permission from the Trane Co., La Crosse, WI.

where Q = rate of heat transfer per ft^2 of surface
Δt = temperature difference
R = resistance

Thermal resistance (R) is related to the K value as follows:

$$R = \frac{d}{K}$$

(11-2)

where d = thickness of material
K = thermal conductivity

For mediums of several materials, the total thermal resistance (R_{total}) is simply the sum of the individual components:

$$R_{total} = R_1 + R_2 + \cdots + R_{N-1} + R_N$$

(11-3)

where R_i = thermal resistance of the ith component, $i = 1, 2, 3, \ldots, N$.

The overall conductance (U) of the total structure is

$$U = \frac{1}{R_{total}}$$

(11-4)

It is important to note that only the resistances are additive. *It is wrong to convert each R to a U value and then add.*

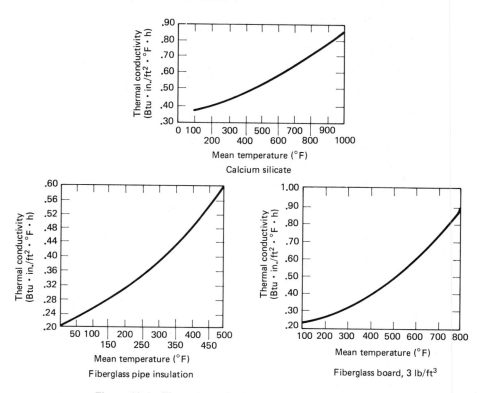

Figure 11-1 Thermal conductivities at varying temperatures.

The final relationships are as follows:

$$Q = \frac{1}{R_1 + R_2 + \cdots + R_{N-1} + R_N} \Delta t \tag{11-5}$$

$$Q = \frac{1}{R_{\text{total}}} \Delta t \tag{11-6}$$

$$Q = U \Delta t \tag{11-7}$$

$$Q_{\text{total}} = UA \Delta t \tag{11-8}$$

where Q_{total} = rate of heat transfer for total surface area involved
  A = area of heat transfer surface

The final concept to be considered here is that of the surface film coefficient, which is the amount of heat transferred from a surface to air or from air to a surface per square foot of surface for 1 deg of temperature difference. Surface film coefficients are usually specified in terms of the surface resistance, as shown in Table 11-2.

As an example of these calculations, assume we have a $\frac{1}{2}$ in. mild steel tank storing a hot fluid as demonstrated in Figure 11-2. The fluid is heated to 200°F, while ambient air is 70°F. What is the heat loss for the uninsulated tank? Ignore the heat loss to the ground. (Actually, transfer to the ground is a rather significant heat loss, but we will ignore it here.) The tank is 10 ft in diameter and 10 ft tall.

There are two primary sources of heat loss shown in Figure 11-2. Q_1 is the loss through the walls of the tank, and Q_2 is the loss through the roof. Although the surface film coefficient will vary somewhat for vertical heat flow vs. horizontal, we will ignore that here, so R_{total} is equal for the roof and walls:

$$R_{\text{total}} = R_{\text{tank}} + R_{\text{surface coefficient}}$$

The resistance of the tank itself is quite small and is usually ignored, but we include it below for demonstration purposes; note how small it is compared to the surface resistance:

$$R_{\text{tank}} = \frac{d}{K} = \frac{.5}{314.4} = .0016 \frac{\text{h} \cdot \text{ft}^2 \cdot {}^\circ\text{F}}{\text{Btu}}$$

$$R_{\text{surface coef.}} = .50 \text{ (Table 11-2, assuming a surface temp. of 120°F)*}$$

$$U = \frac{1}{R_{\text{surface}} + R_{\text{tank}}} = \frac{1}{.5016} = 1.99 \frac{\text{Btu}}{\text{ft}^2 \cdot {}^\circ\text{F} \cdot \text{h}}$$

$$A = \text{area} = \pi DH + \pi r^2 = \pi(10 \text{ ft})(10 \text{ ft}) + \pi(5 \text{ ft})^2 = 392.7 \text{ ft}^2$$

$$Q_{\text{total}} = UA \Delta t = \frac{1.99 \text{ Btu}}{\text{ft}^2 \cdot {}^\circ\text{F} \cdot \text{h}} (392.7 \text{ ft}^2)(200 - 70 {}^\circ\text{F})$$

$$= 101,600 \frac{\text{Btu}}{\text{h}}$$

*We will show you how to check this assumption shortly.

TABLE 11-2 SURFACE FILM COEFFICIENTS, R_s VALUES[a] (h•ft^2•°F/Btu)

$t_s - t_a$ (°F)[b]	Still air		
	Plain fabric, dull metal, $\epsilon = .95$	Aluminum, $\epsilon = .2$	Stainless steel, $\epsilon = .4$
10	.53	.90	.81
25	.52	.88	.79
50	.50	.86	.76
75	.48	.84	.75
100	.46	.80	.72
	With wind velocities		
Wind velocity (mph)			
5	.35	.41	.40
10	.30	.35	.34
20	.24	.28	.27

[a]For heat loss calculations, the effect of R_s is small compared to R_1, so the accuracy of R_s is not critical. For surface temperature calculations, R_s is the controlling factor and is therefore quite critical. The values presented are commonly used values for piping and flat surfaces.

[b]Note that t_s = surface temperature. Knowing the surface temperature requires measurement or calculation through the concept of thermal equilibrium, which will be discussed.

Source: Courtesy of Manville Corp.

To show the impact insulation can have on that, let's assume that aluminum-jacketed fiberglass insulation only 1 in. thick covered the tank. The new heat loss would be

$$R_{insulation} = \frac{d}{K} = \frac{1}{.25} \qquad \text{(K chosen for fiberglass at a mean temperature of around 140°F)}$$

$$R_{insulation} = 4$$

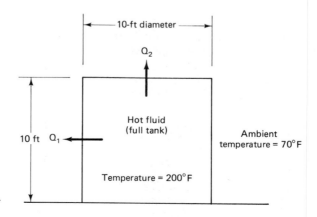

Figure 11-2 Tank storage of hot fluid.

$$U = \frac{1}{R_{\text{tank}} + R_{\text{insulation}} + R_{\text{surface}}} = \frac{1}{.0016 + 4 + .88}$$

(R_{surface} is now chosen for aluminum, assuming a surface temperature of around 95°F.)

$$U = \frac{1}{4.8816} = .205$$

$$Q_{\text{total}} = UA\,\Delta t = \frac{.205\ \text{Btu}}{\text{ft}^2\cdot°\text{F}}(392.7\ \text{ft}^2)(130°\text{F}) = 10{,}465\ \frac{\text{Btu}}{\text{h}}$$

$$\text{savings} = 101{,}796 - 10{,}465 = 91{,}331$$

$$\%\ \text{savings} = \frac{91{,}331}{101{,}796} = \underline{\underline{89.7\%}}$$

Note that this is almost a 90% saving for only 1 in. of insulation. For 2 in., the heat loss would be

$$Q_{\text{total}} = UA\,\Delta t = \frac{1}{8.8816}(392.7)(130) = 5748\ \frac{\text{Btu}}{\text{h}}$$

$$\text{savings} = 101{,}796 - 5748 = 96{,}048\ \frac{\text{Btu}}{\text{h}}$$

$$\%\ \text{savings} = \frac{96{,}048}{101{,}796} = \underline{\underline{94.4\%}}$$

This demonstrates a characteristic of insulation. The first increment yields by far the largest savings, and each additional increment adds to the savings but at a rapidly decreasing rate. At some point, then, it becomes uneconomical to add any additional insulation. This will be demonstrated later.

For pipes, the calculations become a bit more difficult, because the heat flow is in a radial direction away from the pipe through the insulation to a larger surface area. Because of the dispersion of the heat over a larger surface, the "effect" is to increase the insulation thickness.

This effect is manifested in the calculations through a concept known as equivalent thickness d':

$$d' = r_2 \ln \frac{r_2}{r_1} \tag{11-9}$$

where d' = equivalent thickness
r_1 = outside radius of pipe
r_2 = outside radius of pipe plus insulation

For example, a 6-in. nominal size pipe with 2 in. of insulation would have an equivalent thickness of

$$d' = r_2 \ln \frac{r_2}{r_1}$$

$$= 5.313 \ln \frac{5.313}{3.313} = 2.51$$

where r_1 = 3.313 from Table 11-3
r_2 = 3.313 + 2.000 = 5.313

TABLE 11-3 NOMINAL PIPE SIZE VS. OUTSIDE RADIUS

Nominal pipe size	Outside radius	Nominal pipe size	Outside radius
$\frac{1}{2}$	.420	7	3.813
$\frac{3}{4}$	.525	8	4.313
1	.658	9	4.813
$1\frac{1}{4}$	.830	10	5.375
$1\frac{1}{2}$	.950	11	5.875
2	1.188	12	6.375
$2\frac{1}{2}$	1.438	14	7.000
3	1.750	16	8.000
$3\frac{1}{2}$	2.000	18	9.000
4	2.250	20	10.000
$4\frac{1}{2}$	2.500	24	12.000
5	2.781	30	15.000
6	3.313		

The equivalent thickness can now be used in the normal equations to determine the effective R value, etc.

For example, the same 6-in. pipe carrying fluid at 200°F and insulated by 2 in. of aluminum-jacketed fiberglass in a 70°F ambient area would have a heat loss of

$$R_{\text{insulation}} = \frac{d'}{K} = \frac{2.51}{.25} = 10.04$$

$$Q_{\text{total}} = UA\,\Delta t$$

$$= \left(\frac{1}{10.44 \ + \ .88}\right) A(200-70)$$

$$= \frac{11.9 \text{ Btu}}{\text{ft}^2\cdot\text{h}}$$

Since each linear foot of 6-in. pipe has a surface area of

$$\frac{[3.313(2)\,\text{in.}]\pi(1\,\text{ft})}{12\,\text{in./ft}} = 1.735\,\text{ft}^2$$

The heat loss per 1-ft length of pipe is

$$\frac{(11.9\,\text{Btu})}{\text{ft}^2\cdot\text{h}}\frac{(1.735\,\text{ft}^2)}{\text{ft}} = 20.646\ \frac{\text{Btu}}{\text{ft}\cdot\text{h}}$$

The concept of thermal equilibrium is very important in many types of calculations including the checking of surface temperature assumptions deomonstrated earlier. Thermal equilibrium simply says that the total heat flow through a system is equal to the heat flow through any part of the system. For example, a system as shown in Figure 11-3 consists of a wall, two layers of insulation, and an outside surface film. Thermal equilibrium states that*

$$Q_{total} = \frac{t_h - t_a}{R_{i1} + R_{i2} + R_s} = \frac{t_h - t_{i1}}{R_{i1}} = \frac{t_{i1} - t_{i2}}{R_{i2}} = \frac{t_{i2} - t_a}{R_s}$$

$$= \frac{t_{i1} - t_a}{R_{i2} + R_s} \tag{11-10}$$

By using these equations, any surface temperature or other unknown quantity can be calculated if the heat flow and all other quantities are known.

The tank example demonstrated in Figure 11-2 assumed a surface temperature of 95°F. To check this assumption, the total heat loss is known and can be set equal to one of the expressions as shown in Equation 11-10. That is,

$$\frac{Q_{total}}{ft^2} = \left(\frac{10{,}500 \text{ Btu}}{h}\right)\left(\frac{1}{392.7 \text{ ft}^2}\right) = \frac{t_h - t_s}{R_{tank} + R_{insul.}}$$

$$= \frac{200 - t_s}{.0016 + 4}$$

$$\underline{t_s = 93.0°F}$$

As a further check,

$$Q_{total} = \left(\frac{t_s - t_a}{R_s}\right)(392.7 \text{ ft}^2)$$

$$= \left(\frac{93.0°F - 70°F}{.88}\right)(392.7 \text{ ft}^2)$$

$$= 10{,}300 \frac{\text{Btu}}{h}$$

Our assumption was a surface temperature of 95°F. This is close enough.

These calculations are extremely robust in that they can be used for many different objectives. For example, suppose the objective of insulation were to provide personnel protection so the surface temperature could not be higher than, say, 110°F. The energy manager could take the appropriate expressions as in Equation 11-10 and set the surface temperature equal to 110°F. By back-calculating, the required amount of insulation thickness could be determined.

Suppose the purpose were to prevent condensation from forming on cold pipes. Then the energy manager could determine the dew point of ambient air, set the surface temperature equal to that, and back-solve for the insulation thickness.

In both cases, the energy manager would probably round off to the next highest insulation thickness or even increment up one thickness as a safety measure.

For energy management purposes, it is usually necessary to calculate the heat

---

*For purposes here, the inside surface film resistance is assumed to be 0. If not, there would be another set of expressions involving R_s.

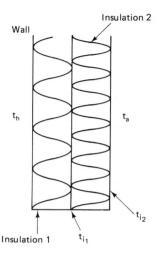

Figure 11-3   Insulation system.

loss (gain) over a year instead of simply by air as already shown. To do this, it is only necessary to sum over the hours.

For example, suppose the pipe in the previous example carried hot fluid 24h/day for 365 days/year. Further, assume that the unit generating the hot fluid is 80% efficient using natural gas costing $5.00/$10^6$ Btu. The total heat loss for a year for the uninsulated system is:

$$Q_{total} = UA\, \Delta t \qquad \text{(assume } T_S \approx 120°\text{F)}$$

$$= \left(\frac{1}{.50}\right)(1)\,(130)$$

$$= 260\ \frac{\text{Btu}}{\text{ft}^2\text{·h}}$$

$$Q_{total(\$)} = \left(\frac{260\ \text{Btu}}{\text{ft}^2\text{·h}}\right)\left(\frac{1}{.8}\right)\left(\frac{\$5.00}{10^6\ \text{Btu}}\right)\left(\frac{8760\ \text{h}}{\text{year}}\right)$$

$$= \$14.24/\text{ft}^2$$

Savings for insulating the pipe would be

$$\left(\frac{260\ \text{Btu}}{\text{ft}^2\text{·h}} - \frac{11.9\ \text{Btu}}{\text{ft}^2\text{·h}}\right)\left(\frac{1}{.8}\right)\left(\frac{\$5.00}{10^6\ \text{Btu}}\right)\left(\frac{8760\ \text{h}}{\text{year}}\right) = \underline{\underline{\$13.58/\text{ft}^2}}$$

Savings per linear foot of pipe would be

$$\left(\frac{\$13.58}{\text{ft}^2}\right)\left(\frac{1.735\ \text{ft}^2}{\text{ft}}\right) = \underline{\underline{\$23.57/\text{ft}}}$$

Consequently, insulation can be a real money saver.

In dealing with systems exposed to outside conditions, it is often helpful to use the degree day or degree hour concept developed in Chapter 6. They can be used as in the example below.

Assume a building wall has an R value of 3 presently. The temperature inside is kept at 65°F during the winter and 75°F during the summer. The plant operates 365 days/year, 24 h/day. Assuming a heating plant efficiency of .8 and a cooling coefficient of performance (COP) of 2.5, what is the cost of energy loss through the wall? Electricity costs $.08/kwh and gas $6.00/10⁶ Btu. The plant experiences 4000°F heating days and 2000°F cooling days. The total wall area is 1000 ft².

$$
\text{heat loss (\$)} = \left(\frac{4000°\text{F days}}{\text{year}}\right)\left(\frac{1 \text{ Btu}}{3 \text{ ft}^2\cdot\text{h}\cdot°\text{F}}\right)\left(\frac{24\text{ h}}{\text{day}}\right) (100 \text{ ft}^2)
$$

$$
\times \left(\frac{1}{.8}\right)\left(\frac{\$6.00}{10^6 \text{ Btu}}\right)
$$

$$
= \$24.00/\text{ft}^2\cdot\text{year}
$$

$$
\text{cooling gain (\$)} = \left(\frac{2000°\text{F days}}{\text{year}}\right)\left(\frac{1 \text{ Btu}}{3 \text{ ft}^2\cdot\text{h}\cdot°\text{F}}\right)\left(\frac{24\text{ h}}{\text{day}}\right) (100 \text{ ft}^2)
$$

$$
= \times \left(\frac{1}{2.5}\right)\left(\frac{\text{kwh}}{3412 \text{ Btu}}\right)\left(\frac{\$.08}{\text{kwh}}\right)
$$

$$
= \$15.00/\text{ft}^2\cdot\text{year}
$$

$$
\text{total energy loss} = \$15.00 + \$24.00 = \$39.00/\text{ft}^2\cdot\text{year}
$$

If the plant experienced a night setback, then it would be necessary to go to °F hours calculated through manipulation of the base corresponding to the hours of setback. For example, if the setback were to 55°F, the heat loss would be calculated according to an inside temperature of 55°F instead of 65°F for the respective hours of setback.

11.2 INSULATION TYPES

Before selecting the proper type of insulation for a particular application, it is important to understand the properties or parameters involved. In this section, the important properties are discussed first, followed by a discussion of the different types of insulation.

11.2.1 Insulation Properties

Some of the more important insulation properties include the following:

- *Cell structure.* Cell structures are either open or closed. A closed cell is relatively impervious to moisture, especially in a moderate environment, so no additional moisture barrier may be needed. Open cells pass moisture freely and therefore probably require vapor barriers. On extremely cold applications where a lot of condensation occurs, a vapor barrier is probably required regardless of cell structure.

- *Temperature use.* Various insulating materials react to extreme temperatures in different ways. In some cases, high temperatures might destroy binders, rendering the insulation useless. All insulation materials then have temperature ranges for which they are recommended. Usually the restriction occurs on the upper end rather than the cold end.

- *Thermal conductivity (K).* As mentioned earlier, K values vary with the temperature—sometimes significantly. The energy manager must be familiar with the different types of insulation, their K values, and how temperature affects the K values. In all cases, the K chosen should be that for the mean temperature (t_m) experienced by the insulation:

$$t_m = \frac{t_h + t_s}{2}$$

- *Fire hazard.* Fire hazard ratings measure the product's contribution to a fire in flame spread and smoke development. It is measured on a frame spread-smoke spread scale where 100/100 is that of red oak.

- *Forms.* Insulation can be made available in a number of different forms. Flexible blankets, batts, rigid board, blocks, and pipe half sections are some of the more popular ones. Insulation is also available in a number of sizes and thicknesses. For example, fiberglass batt insulation with a Kraft paper vapor barrier is available in 15- and 23-in. widths for thicknesses of 3 and 6 in.

11.2.2 Insulating Materials

Some of the more popular types of materials with a discussion of their properties are given here. Details are summarized in Table 11–4.

- *Mineral fiber-rock wool.* Mineral fiber insulation is made from molten rock. It is fairly impervious to heat and so can be used in relatively high temperatures (see Table 11.4).

- *Fiberglass.* Probably the most popular type of insulation, fiberglass can be obtained in blankets, batts, boards, and pipe covering. Although organic binders are frequently used, limiting temperature ranges somewhat, cell structure is

TABLE 11-4 INDUSTRIAL INSULATION

Insulation type and form[a]	Temp. range (°F)	Thermal conductivity [Btu·in./h·ft²·°F at T_m (°F)]			Compressive strength (psi) at % deformation	Fire hazard classification or flame spread-smoke developed	Cell structure (permeability and moisture absorption)
		75	200	500			
Calcium silicate blocks, shapes, and P/C	To 1500	.37	.41	.53	100–250 at 5%	Noncombustible	Open cell
Glass fiber blankets	To 1200	.24–.31	.32–.49	.43–.73	.02–3.5 at 10%	Noncombustible to 25/50	Open cell
Glass fiber boards	To 1000	.22	.28	.51–.61			
Glass fiber pipe covering	To 850	.23	.30	.62			
Mineral fiber blocks and P/C	To 1900	.23–.34	.28–.39	.45–.82	1–18 at 10%	Noncombustible to 25/50	Open cell
Cellular glass blocks and P/C	−450 to 900	.38	.45	.72	100 at 5%	Noncombustible	Closed cell
Expanded perlite blocks, shapes, P/C	To 1500	—	.46	.63	90 at 5%	Noncombustible	Open cell
Urethane foam blocks and P/C	(−100 to −450) to 225	.16–.18	—	—	16–75 at 10%	25–75 to 140–400	95% closed cell
Isocyanurate foam blocks and P/C	To 350	.15	—	—	17–25 at 10%	25–55 to 100	93%closed cell
Phenolic foam P/C	−40 to 250	.23	—	—	13–22 at 10%	25/50	Open cell
Elastomeric closed cell sheets and P/C	−40 to 220	.25–.27	—	—	40 at 10%	25–75 to 115–490	Closed cell
MIN–K blocks and blankets	To 1800	.19–.21	.20–.23	.21–.24	100–190 at 8%	Noncombustible	Open cell
Ceramic fiber blankets	To 2600	—	—	.38–.54	.5–1 at 10%	Noncombustible	Open cell

[a] P/C means pipe covering.

Source: Wayne C. Turner, Senior ed., *Energy Management Handbook*, © 1982. Reproduced by permission of Wiley-Interscience, New York.

such that the limitations can sometimes be exceeded and still have acceptable results.

- *Foams.* Several foam types of insulation are available; some types have problems meeting fire hazard classifications but have very good K values. Others might meet the fire hazard requirements but not offer very good K values. Foams are particularly applicable to cold applications.
- *Calcium silicate.* A very popular type of insulation for high-temperature use, calcium silicate is spun from lime and silica. It is extremely durable and offers a high thermal resistance.
- *Refractories-ceramic fiber.* An alumina-silica product, ceramic fibers are available in blankets or felts that can be used alone or retroadded to existing fire brick.
- *Refractories-fire brick.* Fire bricks are made for high-temperature applications. Made of a refractory clay with organic binders which are burned out during manufacture, they offer good thermal resistance and low storage of heat.
- *Others.* Other types of insulation include cellular glass, perlite, and diatomacious earth. Each has its advantages and disadvantages with which the energy manager must become familiar.

11.3 ECONOMIC THICKNESSES

As might be expected, insulation applications do have optimum thicknesses that can be calculated using the good engineering economy discussed in Chapter 4. Consider Figure 11-4.

As more and more insulation is added, the first cost of material and installation goes up, sometimes in discrete jumps as multiple layers are required. The cost of lost energy, on the other hand, goes down, but at a decreasing rate. At some point, then, the total cost, which is a sum of the lost energy cost and material cost, reaches a minimum point. That is the economic thickness.

It is important to note that the total cost curve is relatively flat in the immediate neighborhood of the economic thickness. This means the energy manager need only get close to the optimum. A small deviation either way won't affect the resulting annual equivalent cost very much.

To determine this economic thickness, the energy manager needs to construct cash flow diagrams for the different alternative thicknesses and calculate the annual equivalent cost for each increment. Since the cash flows include future fuel costs, careful handling of inflation is required.

Many simplified charts, graphs, tables, and computer programs have been developed to help determine optimum thicknesses. The Bibliography includes some of these, but it is pretty easy to develop a program that will quickly calculate equivalent

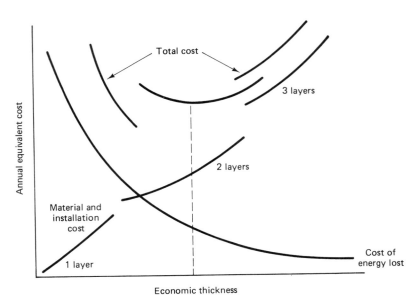

Figure 11-4 Economics and insulation thicknesses.

annual cost for each increment. Figure 11-5 is a printout resulting from a program written at Oklahoma State University. Note in this printout how flat the total cost curve is in the neighborhood of the optimum thickness.

11.4 SUMMARY

Knowledge of insulation is vital for any energy manager. The different types of insulation, their respective advantages and disadvantages, heat loss-gain calculations, and insulation economics are all required information. In this chapter we start that knowledge process.

First, heat loss-gain calculations are reviewed. The concepts of thermal conductivity (K), thermal resistance (R), conductance, and the different types of heat transfer are all reviewed, leading to the heat loss equation:

$$Q = UA \, \Delta T$$

Examples are provided to illustrate the concepts.

The handling of pipes is also covered by demonstrating that the effective thickness (d') of pipe insulation is greater than the actual thickness due to the larger surface area. The equivalent thickness is calculated as

ECONOMIC THICKNESS DETERMINATION

DEPARTMENT OF INDUSTRIAL ENGINEERING AND MANAGEMENT
OKLAHOMA STATE UNIVERSITY

INPUT PARAMETERS

FIRM: PLASTICS, INC.   CONTACT: S. A. PARKER

SYSTEM: PLANT WALLS   DATE: OCTOBER 10, 198–

INSULATION: FIBERGLASS BATTS   K-VALUE: .3600 BTU · IN/H · FT² · °F

AFTER-TAX MARR: 18.0%   GENERAL INFLATION RATE: 10.0%

FUEL ESCALATION RATE: 20.0%   PRESENT R-VALUE: 1.427 HR · FT² · °F/BTU

HEATING DEGREE-HOURS: 112392.0   COOLING DEGREE-HOURS: 153600.0

HEAT PLANT EFFICIENCY: 75.0%   COOLING PLANT EFFICIENCY: 210.0%

COST PER MMBTU OF HEAT: $3700   COST PER MMBTU OF COOLING: $4900

INCREMENTAL TAX RATE: 48.0%   AVAILABLE TAX CREDIT: .0%

PROJECT LIFE: 10 YEARS   DEPRECIATION LIFE: 5 YEARS

THICKNESS CALCULATIONS

INSULATION THICKNESS (IN.)	INSTALLED COST ($/FT²)	ANNUAL ENERGY LOSS OR GAIN (BTU/FT²)	NPV OF SAVINGS ($/FT²)	ANNUALIZE COST ($/FT²)
.00	.00	156271.1	.00	.493
1.00	1.31	53034.6	.52	.378
2.00	1.53	31936.6	.64	.351
3.00	1.65	22847.5	.68	.342
4.00	1.84	17785.7	.62	.355
5.00	2.01	14560.0	.54	.372
6.00	2.15	12324.7	.48	.387
7.00	3.27	10684.4	−.31	.563
8.00	3.43	9429.4	−.41	.584
9.00	3.57	8438.3	−.52	.608

Figure 11–5 Computer printout of insulation economic thickness problem.

$$d' = r_2 \ln \frac{r_2}{r_1}$$

where d' = equivalent thickness
r_1 = outside radius of pipe
r_2 = outside radius of pipe plus insulation

Thermal equilibrium states that once equilibrium is established, the rate of heat loss in any subsection of the insulating system is equal to the total rate of heat loss.

For a system of two insulating media, this leads to

$$Q_{\text{total}} = \frac{t_h - t_a}{R_{i1} + R_{i2} + R_s} = \frac{t_h - t_{i1}}{R_{i1}} = \frac{t_{i1} - t_{i2}}{R_{i2}} = \frac{t_{i2} - t_a}{R_s}$$

$$= \frac{t_{i1} - t_{ia}}{R_{i2} + R_s}$$

where Q_{total} = total rate of heat flow
  t_h = temperature on "hot" side
  t_a = ambient temperature
  R_{i1} = resistance of first section of insulation
  R_{i2} = resistance of second section of insulation
  R_s = resistance of outside surface
  t_{i1} = temperature of inside surface of insulation 2 or outside surface temperature of insulation 1
  t_{i2} = outside surface temperature

Next we discussed how the thermal equilibrium equations could be used to make the necessary calculations for personnel safety (surface temperature safe to touch) and moisture control (surface temperature higher than ambient air dew point).

Then, we demonstrated how heat loss (gain) can be summed over a year by using hours of exposure or degree days-degree hours.

Next we discussed some important properties of the different insulation types. Properties discussed include cell structure, reaction to temperature, conductivity (K), fire hazard, and forms. Then some of the more popular types of insulation were presented, including mineral fiber, fiberglass, foams, calcium silicate, ceramic fibers, and fire bricks.

Finally, a short discussion was given on the concept of economic thickness. It was discovered that by running life-cycle costs for thickness increments, a nice convex curve results with a rather flat region at the bottom. This means that economic thicknesses can be determined easily and that missing the true optimum point a little does not affect the total cost substantially.

BIBLIOGRAPHY

PARKER, STEVEN A., *Insulate,* working paper, Oklahoma State University, Stillwater, Okla., 1982.

THUMANN, ALBERT, *The Plant Engineer's and Manager's Guide to Energy Conservation,* Van Nostrand Reinhold, New York, 1977.

TURNER, WAYNE C., Senior ed., *Energy Management Handbook,* Wiley-Interscience, New York, 1982.

QUESTIONS

11.1. Give an example of heat transfer by radiation, conduction, and convection.

11.2. Given that infrared heaters heat by radiation, why are they recommended for large open areas or areas with a lot of air infiltration?

11.3. Discuss whether insulation actually stops heat loss or only slows it down.

11.4. Demonstrate why the R value of a metal tank itself is usually ignored and the surface resistance R_s is used.

11.5. If it's necessary to calculate an effective insulation thickness for pipes, why isn't it necessary to do the same for tanks?

11.6. Discuss why the concept of thermal equilibrium is so important.

PROBLEMS

11.1. Given a metal tank 4 ft in diameter, 6 ft long, holding water at 180°F, what is the heat loss per year in Btu? The tank holds hot water all the time and is on a stand so all sides are exposed to ambient conditions at 80°F. If the boiler supplying this hot water is 79% efficient and uses natural gas costing $5.00/$10^6$ Btu, what is the cost of this heat loss? There is no air movement around the tank.

11.2. Ace Manufacturing has a condensate return tank holding pressurized condensate at 20 psig sat. The tank is 2.5 ft in diameter and 4 ft long. Presently, it is not insulated. Management is considering adding 2 in. of aluminum-jacketed fiberglass at an installed cost of $.60/$ft^2$. Calculate the present worth of the proposed investment. The steam is generated by a boiler 78% efficient consuming No. 2 fuel oil at $7.00/$10^6$ Btu. Energy costs will remain constant over the economic life of the insulation of 5 years. Ambient temperature is 70°F. Use R_s = .42 for the uninsulated tank. The tank is utilized 8000 h/year.

11.3. Your plant has 500 ft of uninsulated hot water lines carrying water at 180°F. The pipes are 4 in. in nominal diameter. You decide to insulate these with 2-in. calcium silicate snap-on insulation at $1.00/$ft^2$ installed cost. What is the savings in dollars and Btu if the boiler supplying the hot water consumes natural gas at $6.00/$10^6$ Btu and is 80% efficient? Ambient air is 80°F, and the lines are active 8760 h/year.

11.4. Given a wall constructed as shown in Figure P11–1, what is the cost of heat loss and heat gain for a year per ft^2? Heating degree days are 4000°F days, while cooling degree days are 2000°F days. Heating is by gas with a unit efficiency of .7. Gas costs $6.00/$10^6$ Btu.

Cooling is by electricity at $.06/kwh (ignore demand costs), and cooling plant efficiency is 2.5 seasonal.

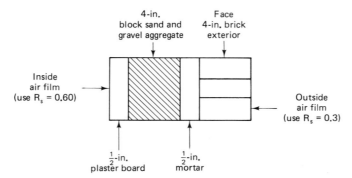

11.5. Given a 6-in. pipe carrying chilled water at 40°F in an atmosphere with a temperature of 90°F and a dew point of 85°F, how much fiberglass insulation with Kraft paper jacket is necessary to prevent condensation on the pipes?

Chapter 12

. . . .

Process
Energy Management

12.0 INTRODUCTION

In many organizations, energy management is simply a matter of managing the energy required for lighting and space conditioning. In many others, however, energy management is much more complex and could involve industrial insulation, complex combustion monitoring, unique steam distribution problems, significant amounts of waste heat, etc. Typical institutions offering such large problems (and subsequently opportunities) include hospitals, prisons, and, most notably, industry.

In this book we have presented material that enables the energy manager to perform necessary activities in all these areas, but process energy management is so important, we thought a chapter devoted to it alone would be helpful. Most of the material covered and the examples presented draw from the other chapters.

The energy manager must be careful in process energy management. Processes can be quite complex, so a full understanding of the entire process is necessary. Otherwise suboptimization can occur, or, even worse, harmful impacts can be created. For example, small changes in heat treatment temperatures or atmosphere can sometimes dramatically affect the product quality or subsequent workability.

In this chapter we first present a suggested procedure for process energy management with some short examples. Then some popular process activities for better energy management are outlined. Finally, some examples of process energy management are provided.

12.1 STEPS FOR IMPROVEMENT

Readers who have studied work simplification and improvement may remember the suggested order of changes as (1) eliminate; (2) combine; (3) change sequence, person, place, or equipment; and (4) improve.

The same order of change is appropriate for process energy management, but as mentioned earlier, the analyst must understand the entire system and the cascading impacts that changes might effect. In terms of energy management, examples of the preceding changes include the following:

- *Eliminate.* Does that cooling water really need to be there? Sometimes process cooling water really isn't necessary, saving pumping and chilling costs. Is the paint oven really necessary? Newer paints sometimes will air dry quite well, and paint oven costs can be substantial.

- *Combine.* Quite often machining operations can be combined with jig and fixture modifications or changes in equipment. This saves energy in the second machine (and maybe in more machines) as well as material handling and possibly process storage energy. Sometimes combining processes also saves the energy necessary to bring the material back to a required workability.

- *Change sequence, person, place, or equipment.* Equipment changes often offer substantial energy savings as the newer equipment may be more energy efficient. For example, newer electric welders are considerably more energy efficient than older ones. Changing sequences, persons, or place can offer energy savings as the place may be more appropriate, the person more skillful, and the sequence better in terms of energy consumption. For example, bringing rework back to the *person* with the skill and *place* with the correct equipment can save energy.

- *Improve.* Most energy management work today is in improving because the capital expenditure required is often minimized. Examples include reducing excess air for combustion to a minimum, reducing temperatures to the minimum required (don't forget chilling—maybe the freezer temperature can be increased a few degrees), and removing excess lighting. Improving does sometimes require large amounts of capital. For example, insulation improvements can be expensive, but energy savings can be large, and there can be improved product quality.

12.2 TWENTY-FIVE POPULAR ENERGY MANAGEMENT OPPORTUNITIES

Through our combined energy management experiences with around 200 manufacturing plants and a review of the literature, we have found that there are some very popular energy management opportunities (EMOs) that tend to appear time after time. The astute energy manager will become familiar with these opportunities and be ready to apply them (*as well as others*) in energy management work.

Twenty-five of these popular changes are summarized below. Most are process modifications, but we decided to show all popular changes, so a few are lighting and space conditioning oriented. The order of listing does not imply anything, for the order would change if one were listing by frequency of occurrence, amount of savings, financial return on investment, etc.

1. *Switch to energy-efficient lights.* Switch existing lighting to the energy-efficient ones, such as 34–35-w energy-efficient fluorescent bulbs for convential 40-w ones. Also lower wattages.

2. *Switch to energy-efficient light sources.* Change to more efficient sources usually requiring fixture changes. Change from incandescent lights to fluorescent or from mercury vapor or fluorescent to high-pressure sodium for in-plant lighting, a very popular conversion.

3. *Use night setback-setup.* Turn back (up) temperatures at night when needs are reduced. Examples include large ovens that can't be turned off, large refrigeration units where night operations involve less infiltration (people going in and out), and space conditioning.

4. *Turn off equipment.* Turn off exhaust fans, ovens, motors, or any equipment when not needed.

5. *Move air compressor intake to cooler locations.* Move air intakes from hot equipment rooms to cooler (often outside) locations. Efficiency improvements are large and paybacks attractive.

6. *Eliminate leaks in steam and compressed air.* Steam and compressed air leaks are very expensive and should be fixed. Technology exists for repairing leaks without shutting the equipment down. Night audits (when noise is minimized) often turn up large amounts of these leaks.

7. *Control excess air.* As shown in earlier discussions, careful control of combustion air can lead to dramatic energy savings

8. *Optimize plant power factor.* Depending on the utility billing schedule and the company's power factor, dramatic savings may be available through power factor improvement.

9. *Insulate bare tanks, vessels, lines, and process equipment.* Big savings are often available through insulation of process lines and tanks. Condensate return lines and tanks are often not insulated.

10. *Install storm windows, doors, and weather stripping.* Though often difficult to justify economically, big savings are sometimes available. This is especially true for large glass exposures in cold climates.

11. *Use energy-efficient electric motors.* When replacement is necessary or for new applications, energy-efficient motors can usually be justified.

12. *Preheat combustion air.* Recuperators can save large amounts of energy and money. Sometimes they are quite cost effective.

13. *Reduce the pressure of compressed air and steam.* Sometimes pressures are overdesigned, and a reduction won't harm the process. In such cases, large savings are possible.

14. *Insulate walls, ceilings, roofs, and doors.* Industrial plants are frequently poorly insulated. Sometimes insulation in dropped ceilings, on roofs or walls, and doors can be cost-justified.

15. *Recover heat from air compressor.* Larger air compressors reject large amounts of heat through air or water cooling. Proper design can allow the use of this waste heat for space conditioning in the winter and its exhaust in warm weather. Sometimes the payback is very attractive.

16. *Insulate dock doors.* Plastic strips, dock bumpers, vestibules, or air screens all help block infiltration through large dock doors. If the space is heated and/or air-conditioned, savings can be very large.

17. *Install economizers on air conditioners.* In some areas of the country, economizers can be very attractive. They allow the optimum use of outside air in air conditioning. Sometimes outside air can be used and the air conditioner turned off.

18. *Use radiant heat.* Sometimes infrared heaters can be used to spot-heat rather than heat entire areas. Infrared heat (like the sun) warms objects and people but not space. The payback can be very attractive.

19. *Return steam condensate to the boiler.* Returning hot condensate can yield dramatic savings in energy, water, and water conditioning costs. Return lines should probably be insulated.

20. *Change product design to reduce energy requirements.* Product redesign can often reduce the energy necessary in heat treating, cleaning, coating, painting, etc.

21. *Explore waste heat recovery for space exhaust systems.* Large amounts of exhaust in buildings that are heated and/or air-conditioned offer potential for waste heat recovery.

22. *Install devices to improve heat transfer in boilers.* Turbulators and other devices designed to reap more energy out of the combustion process are often very cost effective.

23. *Reschedule operations to reduce peak demand.* Sometimes simple changes in equipment scheduling can dramatically reduce demand charges.

24. *Cover open heated tanks.* Covering open heated tanks can often lead to big energy savings. Floating balls, cantilevered tops, and rubber flaps have all been used.

25. *Spot-ventilate or use air filters.* In welding areas or other areas where large amounts of ventilation are required, spot ventilation can often reduce the

amount needed. Also, electrostatic or other types of air filters can sometimes allow reuse of the air. Savings are especially large if the space is heated and/or air-conditioned.

12.3 EXAMPLES

In this section, some examples of abbreviated studies of process energy improvements are presented. They are given as examples and not actual calculation guidelines.

12.3.1 Recuperator for a Large Brick Kiln

Summary.   A recuperator is a device that preheats the combustion air. Preheating the combustion air increases the burner efficiency. The brick kilns presently use ambient air for the combustion air. The air intake could be modified to draw air from the cooling section of the kiln. This air has the full oxygen content yet has been heated in order to cool the brick. A simple insulated duct could connect the cooling section to the air intake motor. It is recommended that the air drawn from the cooling section be no more than 800°F and the duct be insulated. There might be a need for a filtering system if the new air source has unwanted dust particles. Usually the higher the temperature of the combustion air, the more efficient the combustion process. Unfortunately, when dealing with temperatures above 800°F, there is a risk of the burners becoming too hot, resulting in a shorter burner life. By using air at 800°F and taking into account the heat loss in the duct, the risk of harming the burners is reduced. There might also be a need for additional controls on the air intake motor due to higher air temperatures.

Required Data

Present air intake temperature: 90°F

Proposed air intake temperature: 800°F

Airflow*: 3548 cfm/kiln

Specific heat of air (C_p) at 800°F: .259 Btu/lb·°F

Specific heat of air (C_p) at 90°F: .240 Btu/lb·°F

Density of air (p) at 800°F: .03 lb/ft³

Density of air (p) at 90°F: .07 lb/ft³

Heat loss through ductwork: 30% (using 3-in. hot pipe insulation)

Operating hours: 8760 h/year

Energy cost: $3.30/Mcf

*The airflow is calculated from the proposed burner consumption after excess air control is optimized.

Calculations.   In raising the combustion air temperature, the air mass flow rate must remain constant in order to maintain a correct air-fuel mix:

$$\text{air mass flow rate} = m = (3548 \text{ ft}^3/\text{min}) (.07 \text{ lb}/\text{ft}^3)$$
$$= 248.4 \text{ lb}/\text{min}$$

$$\text{new volume flow rate } (800°\text{F}) = (248.4 \text{ lb}/\text{min}) (1 \text{ ft}^3/.03 \text{ lb})$$
$$= 8280 \text{ cfm}$$

$$\text{savings in Btus} = (\text{air mass flow})(T_2 C_{p1} - T_1 C_{p1})(1 - \text{total heat loss})$$
$$= (248.4 \text{ lb}/\text{min})(800°\text{F} - 90°\text{F}) \times .250^* \text{ Btu}/\text{lb·}°\text{F}$$
$$\times (60 \text{ min}/\text{h})(8760 \text{ h}/\text{year})(2 \text{ kilns})(1 - .30)$$
$$= 32,400 \times 10^6 \text{ Btu}/\text{year}$$

$$\text{savings in Mcf} = (32,400 \times 10^6 \text{ Btu}/\text{year})(1\text{Mcf}/10^6 \text{ Btu})$$
$$= 32,400 \text{ Mcf}/\text{year}$$

$$\text{savings in \$} = (32,400 \text{ Mcf}/\text{year})(\$3.30/\text{Mcf})$$
$$= 107,000/\text{year}$$

Implementation Cost

Length of duct: 75 ft/kiln
Ductwork: 304 stainless steel at $5.45/lb
24-in.-diameter duct of 14-gauge steel: 20 lb/ft
Insulation for ductwork: $10.00/ft
Engineering design cost: $5000
Contingency: $2000

$$\text{implementation cost} = [(75 \text{ ft})(\$5.45/\text{lb})(20 \text{ lb}/\text{ft}) + (75 \text{ ft})(\$10.00/\text{ft})](2 \text{ kilns})$$
$$+ \$5000 + \$2000$$
$$= \$24,850$$

Simple Payback

$$\text{payback} = \frac{\text{implementation cost}}{\text{annual savings}}$$
$$= \frac{\$24,850}{\$107,000/\text{year}}$$
$$= 0.23 \text{ year}$$

12.3.2 Heat Recovery from Compressors to Space-Heat a Warehouse

Summary.   By installing ductwork for an air compressor, we can use the hot air from the existing two compressors to heat a warehouse. For the two compressors (100 and 75 hp), the Btu savings as well as dollar savings will be calculated.

---

* Average specific heat over the temperature range.

Required Data

Average air temperature before compressor (T_1): 90°F
Average air temperature after compressor (T_2): 117°F
Compressor size: 100 and 75 hp
Operation hours: 992 h/year (24 h/day, 5 days/week; 4 h/day, 1 day/week; 8 weeks/year)
Electricity cost: $.034/kwh
Percent load on compressor: 75% (assumed)
Ductwork length: 35 ft each (70 ft for both)

Calculations.   For the 100-hp compressor,

$$
\begin{aligned}
\text{Btu savings} &= (20825 \text{ ft}^3/\text{min})(.075 \text{ lbm}/\text{ft}^3)(.24 \text{ Btu}/\text{lbm·°F})(117 - 90°F) \\
&\quad \times (60 \text{ min}/\text{h})(992 \text{ h}/\text{year}) \\
&= 602.4 \times 10^6 \text{ Btu}/\text{year} \\
&= (602.4 \times 10^6 \text{ Btu}/\text{year})(.75 \text{ load}) \\
&= (451.8 \times 10^6 \text{ Btu}/\text{year})
\end{aligned}
$$

For the 75-hp compressor, we'll assume that the savings will be about 75% of the 100-hp compresor, so we have

$$
\begin{aligned}
\text{Btu savings} &= (451.8 \times 10^6 \text{ Btu}/\text{h})(.75) \\
&= 338.85 \times 10^6 \text{ Btu}/\text{h} \\
\text{Btus saved} &= (451.8 \times 10^6 + 338.85 \times 10^6 \text{ Btu}/\text{h}) \\
&= 790.65 \times 10^6 \text{ Btu}/\text{h} \\
\text{kwh saved} &= (790.65 \times 10^6/\text{year})(1 \text{ kwh}/3412 \text{ Btu}) \\
&= 231,726.26 \text{ kwh} \\
\$ \text{ saved} &= (231,726.26 \text{ kwh})(\$.034/\text{kwh}) \\
&= \$7878.70/\text{year}
\end{aligned}
$$

Implementation Cost

Material cost: For the air intake ductwork, an insulated flexible duct with vinyl-coated spring steel (or aluminim) which costs about $2.90/linear foot and also two dampers at $30 each for $60 can be used.
Labor cost: For the installation of the duct, two persons may be employed at $20/h for about 8 h.

$$
\begin{aligned}
\text{total cost} &= \text{material cost} + \text{labor cost} \\
&= (\text{duct cost}/\text{linear ft})(\text{total linear ft}) + (2 \text{ dampers}) \\
&\quad \times (\$30/\text{damper}) + (\text{number of laborers})(\text{number of h worked}) \\
&\quad (\text{wage}/\text{h}) \\
&= (\$2.90/\text{linear ft})(70 \text{ ft}) + (\$30/\text{damper})(2 \text{ dampers}) \\
&\quad + (2 \text{ laborers})(8 \text{ h})(\$20/\text{h}) \\
&= \$583
\end{aligned}
$$

Simple Payback

$$\text{payback} = \frac{\text{cost}}{\text{savings}}$$

$$= \frac{\$583}{\$7878.7} = .075 \text{ year}$$

Note: For this system, an automatic damper might be required so that the temperature of air for space heating will not be too high. The system then can mix cool outside air with the hot compressor exhaust air if needed. The damper is required also so that air may be exhausted in the summertime.

12.3.3 Installation of an Economizer for a Plastic Plant

Summary. This EMO considers the installation of an economizer in a plastic plant. This area is thermally heavy due to the heat being generated by a large injection molder. To counteract this heat generation, the plastic plant must air-condition 9 months (March through November), three shifts a day. The economizer control will read the dry and wet bulb temperatures to determine if the outside air conditioners (temperature and humidity) are more desirable than the present return air. When outside air is more desirable (less enthalpy), the economizer cycle will allow the outside air and inside air to mix in the proper proportions so that the least amount of energy will be expended to get the air to the required conditions. It was assumed that the return air was at 78°F and 50% relative humidity (a conservative estimate), which corresponds to an enthalpy reading of 30 Btu/lbm (air). Btu savings occur when the enthalpy of the outside air is less than the return air [30 Btu/1 lb (air)]. The savings can be easily calculated in this manner by using bin data,* which give the number of hours of weather experience in a month at a given temperature range.

Required Data

Electrical energy cost: $.034/kwh
Return air: 78°F, 50% relative humidity
Operation time: 3 shifts (24 h/day Monday–Friday; 4 h/day Saturday)

Calculations

$$
\begin{aligned}
\text{savings in energy} &= (\text{total savings of Btu·h/1bm air·year})(\text{cfm of unit}) \\
&\quad (60 \text{ min/h}) \times (\text{density of air})(\tfrac{1}{\text{COP}}) \\
&= (25{,}858.5 \text{ Btu·h/1bm air·year})^\dagger (7120 \text{ ft}^3/\text{min})) \\
&\quad (60 \text{ min/h}) \times (.0769 \text{ 1bm/ft}^3 \text{ air})(\tfrac{1}{3}) \\
&= 283.2 \times 10^6 \text{ Btu/year}
\end{aligned}
$$

$$
\begin{aligned}
\text{savings in dollars} &= (\text{Btu/year})(\text{kwh}/3412 \text{ Btu})(\text{cost/kwh}) \\
&= (283.2 \times 10^6 \text{ Btu/year})(\text{kwh}/3412 \text{ Btu})(\$.034/\text{kwh}) \\
&= \$2821.69
\end{aligned}
$$

---

*See Table 12-1 (obtained from the U.S. Air Force base in the same town).

†See Table 12-2.

Implementation Cost

total cost* = two 7.5-ton economizer units at $1216 each
plus installation cost
= 2($1216) + $600
= $3032

Simple Payback

$$payback = \frac{cost}{savings}$$

$$= \frac{\$3032}{\$2821/year} = 1.07 \text{ years}$$

12.3.4 HPS Relamp of Refrigerated Storage

Summary.   This ECO recommends major relamp of refrigerated storage. The following analysis will show the potential savings of relamping to 70-watt high-pressure sodium (HPS) Lamps from the present 150-watt incandescent lights. The level of high-pressure sodium illumination provided for is based on the existing level of illumination. High-pressure sodium is one of the most efficient high-intensity discharge (HID) source types and has excellent lumen maintenance over life. The golden-white color quality of HPS has been a limiting factor for many interior applications, but improvements in color rendition have made its application for interior lighting quite acceptable even for color-critical areas.

Required Data

Present lighting data:
    Type: incandescent   Initial lumens: 2350
    Size: 150 w (150-w input to fixture)   Life: 2500 h
    Quantity: 68   Cost: $1.10

Proposed lighting data:
    Type: high-pressure sodium   Initial lumens: 5400
    Size: 70 w (88-w input to fixture)   Life: 20,000 h
    Spacing-to-mounting-height ratio: 2.0   Fixture cost: $91.55
                                Light cost: $39.69

Electrical energy cost (including fuel adjustment cost): $.034/kwh

Demand cost: $4.97 (June–October)
                $3.29 (November–May)
                $3.99 (average year-round)

Total area: 9800 ft^2

Mounting height: 8 ft

Hours of operation: 4576 h/year (16 h/day, 5.5 days/week, 52 weeks/year)

Cooling unit coefficient of performance (COP): 2.5 (assumed)

*Obtained from vendor.

TABLE 12.1 MEAN FREQUENCY OF OCCURRENCE OF DRY BULB TEMPERATURE (°F) WITH MEAN COINCIDENT WET BULB TEMPERATURE (°F) FOR EACH DRY BULB TEMPERATURE RANGE

COOLING SEASON

Temperature range (°F)	May Observ./h, Gp 02 to 09	10 to 17	18 to 01	Total observ.	Mean coincident wet bulb (°F)	June Observ./h, Gp 02 to 09	10 to 17	18 to 01	Total observ.	Mean coincident wet bulb (°F)	July Observ./h, Gp 02 to 09	10 to 17	18 to 01	Total observ.	Mean coincident wet bulb (°F)
105–109													0	0	72
100–104							0	0	0	73		7	0	7	72
95–99		1		1	68		8	1	9	74		51	8	59	74
90–94		8	1	9	71		40	7	47	74	1	70	26	97	74
85–89		33	5	38	70	2	64	25	91	72	11	53	43	107	73
80–84	1	58	21	80	68	14	61	53	128	71	34	35	73	142	71
75–79	13	51	46	110	66	50	36	70	156	69	90	20	55	165	70
70–74	53	40	62	155	64	86	17	51	154	67	83	10	34	127	68
65–69	72	29	55	156	61	62	10	24	96	63	26	21	9	37	65
60–64	55	17	32	104	58	20	3	8	31	60	3	0		3	60
55–59	33	9	20	62	53	5	1	1	7	55	0			0	57
50–54	15	2	5	22	48	1	0		1	49					
45–49	5	0	1	6	43										
40–44	1	0		1	40										
35–39	0			0	35										
30–34															
25–29															

Calculations

$$\text{present illumination} = (68 \text{ lamps})(2350 \text{ lm/lamp})$$
$$= 159{,}800 \text{ lm}$$

$$\text{number of HPS lights required} = \frac{159{,}800 \text{ lm}}{5400 \text{ lm/light}}$$
$$= 30 \text{ lights}$$

$$\text{area of lighting/HPS light} = \frac{9800 \text{ ft}^2}{30 \text{ lights}}$$
$$= 326.67 \text{ ft}^2/\text{light}$$

$$\text{spacing requirements} = (326.67 \text{ ft}^2)^{1/2}$$
$$= 18.07 \text{ ft/light}$$

TABLE 12.1 (CONT.)

August					September					October				
Observ./h, Gp 02 to 09	10 to 17	18 to 01	Total observ.	Mean coincident wet bulb (°F)	Observ./h, Gp 02 to 09	10 to 17	18 to 01	Total observ.	Mean coincident wet bulb (°F)	Observ./h, Gp 02 to 09	10 to 17	18 to 01	Total observ.	Mean coincident wet bulb (°F)
	2	0	2	73										
	15	2	17	73										
0	37	7	44	73	0			0	72					
1	63	20	84	73	11	0		11	71		0		0	61
7	64	39	110	72	30	4		34	71		6		6	64
29	38	70	137	71	1	48	13	62	70		14	1	15	66
82	20	64	166	69	7	43	35	85	69		30	5	35	65
84	7	37	128	67	30	42	56	128	67	2	39	14	55	63
37	2	8	47	63	67	33	48	148	65	13	44	37	94	62
7	0	1	8	58	53	18	46	117	62	35	46	50	131	59
1			1	55	45	11	26	82	57	54	34	53	141	55
					27	3	8	38	53	49	21	43	113	51
					7	1	4	12	49	48	8	27	83	47
					3	0		3	46	29	4	13	46	42
										12	1	3	16	38
										4	1	1	6	34
										1	1		2	29
										1	0		1	26

$$\text{spacing-to-mounting-height ratio} = \frac{18.07 \text{ ft}}{8.5 \text{ ft}}$$

$$= 2.1 \quad \text{(this is extremely close to the recommended spacing-to-mounting-height ratio of 2.0 obtained from mfg. catalog)}$$

present kwh = (68 lamps)(150 w/lamp)(4576 h/year)(1 kw/1000 w)
= 46,675.20 kwh/year

proposed kwh = (30 lights)(88 w/light)(4576 h/year)(1 kw/1000 w)
= 12,080.64 kwh/year

savings in kwh = (46,675.20 kwh/year − 12,080.64 kwh/year)

= 34,594.56 kwh/year
= (34,594.56 kwh/year)(3412 Btu/kwh) = 118.04×10^6 Btu/year

TABLE 12.1 (*CONT.*)

HEATING SEASON

Temperature range (°F)	November Observ./h, Gp 02 to 09	10 to 17	18 to 01	November Total observ.	November Mean coincident wet bulb (°F)	December Observ./h, Gp 02 to 09	10 to 17	18 to 01	December Total observ.	December Mean coincident wet bulb (°F)	January Observ./h, Gp 02 to 09	10 to 17	18 to 01	January Total observ.	January Mean coincident wet bulb (°F)
105–109															
100–104															
95–99															
90–94															
85–89						0			0	60					
80–84						0			0	59					
75–79	7			7	60	0			0	55					
70–74	2	25	4	31	60	2	0		2	53	1			1	56
65–69	6	26	14	46	57	7	1		8	51	7	1		8	53
60–64	11	30	22	63	53	1	17	3	21	51	2	13	4	19	52
55–59	19	35	29	83	49	7	27	14	48	49	4	17	8	29	49
50–54	29	38	42	109	45	11	37	24	72	45	7	29	18	54	45
45–49	43	34	40	117	42	23	40	39	102	41	11	34	24	69	41
40–44	39	20	32	91	37	38	38	45	121	38	29	38	43	110	37
35–39	43	14	31	88	33	46	30	41	117	33	36	31	42	109	33
30–34	26	7	14	47	29	49	20	34	103	29	52	28	35	115	29
25–29	10	3	9	22	24	32	14	22	68	25	39	19	26	84	25
20–24	10	1	3	14	20	23	10	14	47	20	24	14	23	61	20
15–19	1	0	0	1	15	8	4	7	19	16	19	7	10	36	16
10–14	1			1	10	7	2	4	13	11	13	7	7	27	11
5–9						3	0	0	3	7	5	3	5	13	6
0–4						0			0	3	6	0		8	2
−5−−1											1	0		1	−2

present kw (demand)　= (68 lamps)(150 w/lamp-month)(1 kw/1000 w)
　　　　　　　　　　= 10.20 kw/month

proposed kw (demand) = (30 lights)(88 w/light-month)(1 kw/1000 w)
　　　　　　　　　　= 2.64 kw/month

savings in kw (demand) = 10.20 kw/month − 2.64 kw/month
　　　　　　　　　　= 7.56 kw/month

present replacement cost　= (68 lamps)($1.10/lamp)(1/2500 h)
　　　　　　　　　　　　(4576 h/year)
　　　　　　　　　　= $136.91/year

TABLE 12.1 (CONT.)

February					March					April					Annual (total all months)				
Observ./h, Gp					Observ./h, Gp					Observ./h, Gp					Observ./h, Gp				
02 to 09	10 to 17	18 to 01	Total observ.	Mean coincident wet bulb (°F)	02 to 09	10 to 17	18 to 01	Total observ.	Mean coincident wet bulb (°F)	02 to 09	10 to 17	18 to 01	Total observ.	Mean coincident wet bulb (°F)	02 to 09	10 to 17	18 to 01	Total observ.	Mean coincident wet bulb (°F)
																2	0	2	73
																23	3	26	73
															0	108	16	124	73
											2		2	65	3	219	57	279	70
	0		0	62		1	0	1	58		10	1	11	66	20	286	128	434	70
	1	0	1	60		4	1	5	61	0	20	5	25	64	85	292	262	639	66
	3	0	3	57	0	10	2	12	59	1	32	15	48	63	269	259	322	850	64
	7	1	8	57	0	15	5	20	56	10	41	33	84	61	398	245	314	957	61
0	12	4	16	54	4	24	13	41	54	31	38	39	108	58	326	219	265	810	58
2	15	7	24	52	10	30	21	61	51	39	33	41	113	54	250	204	219	673	55
7	22	15	44	49	15	30	28	73	48	41	26	37	104	49	207	188	203	598	51
9	29	23	61	45	24	33	36	93	45	41	19	30	90	45	192	195	208	595	46
26	26	33	85	42	36	29	38	103	41	35	13	22	70	42	210	180	209	599	42
32	30	39	101	38	45	28	42	115	39	26	4	12	42	38	223	159	213	595	38
37	32	32	101	33	42	21	28	91	34	12	1	4	17	35	221	132	178	531	34
41	18	29	88	29	36	13	20	69	29	3			3	30	207	85	133	425	29
29	13	22	64	25	21	7	9	37	25	1	1	1	3	26	132	57	89	278	25
23	11	13	47	20	10	2	3	15	20						90	38	56	184	20
12	4	5	21	16	2	1	1	4	16						42	17	25	84	16
5	1	1	7	12	2	0	1	3	11						27	9	13	49	11
1			1	7	1	0		1	8						10	3	5	18	6
					0			0	4						7	0	2	9	2
															1	0		1	−2

proposed replacement cost = (30 lights)($39.69/light)(1/20,000 h)
(4576 h/year)
= $272.43/year

savings in replacement cost = $136.91/year − $272.43/year
= − $135.52/year

Savings in heat removal. The reduction in wattage made possible through HPS relamp will also reduce the amount of heat emitted by the lights that must be removed:

TABLE 12-2   ENTHALPY DATA [BTU•H/LBM (AIR)]

March

Temperature (°F) Dry	Wet	h	$\Delta h_{(30-h)}$	Hours	Btu•hr 1 lbm (air)
77	59	25.8	4.2	12	50.4
72	56	24	6	20	120
67	54	22.6	7.4	41	303.4
62	51	21	9	61	549
57	48	19.2	10.8	73	788.4
52	45	17.6	12.4	93	1153.2
47	41	15.8	14.2	103	1462.6
42	38	14.5	15.5	115	1782.5
					6209.5

April

Temperature (°F) Dry	Wet	h	$\Delta h_{(30-h)}$	Hours	Btu•hr 1 lbm (air)
77	63	28.6	1.4	48	67.2
72	61	27.2	2.8	84	235.2
67	58	25	5	108	540
62	54	22.6	7.4	113	836.2
57	49	19.8	10.2	104	1060.8
52	45	17.6	12.4	90	1116
47	42	16.2	13.8	70	966
42	38	14.5	15.5	42	651
					5472.4

May

Temperature (°F) Dry	Wet	h	$\Delta h_{(30-h)}$	Hours	Btu•hr 1 lbm (air)
77	66	no savings			
72	64	29.4	.6	155	93
67	61	27.2	2.8	156	436.8
62	58	25	5	104	520
57	53	22	8	62	496
52	48	19.2	10.8	22	237.6
47	43	16.8	13.2	6	79.2
42	40	15.2	14.8	1	14.8
					1877.4

TABLE 12.2 *(CONT.)*

June

Temperature (°F)		h	$\Delta h_{(30-h)}$	Hours	Btu·hr / 1 lbm (air)
Dry	Wet				
77	69	no savings			0
72	67	no savings			0
67	63	28.5	1.4	96	134.4
62	60	26.5	3.5	31	108.5
57	55	23.2	6.8	7	47.6
52	49	19.8	10.2	1	10.2
47				None	0
42				None	
					300.7

July

Temperature (°F)		h	$\Delta h_{(30-h)}$	Hours	Btu·hr / 1 lbm (air)
Dry	Wet				
77	70	no savings			0
72	68	no savings			0
67	65	no savings			0
62	60	26.5	3.5	3	10.5
57				None	0
52				None	0
47				None	0
42				None	0
					10.5

August

Temperature (°F)		h	$\Delta h_{(30-h)}$	Hours	Btu·hr / 1 lbm (air)
Dry	Wet				
77	69	no savings			0
72	67	no savings			0
67	63	28.6	1.4	47	65.8
62	58	25	5	8	40
57	55	23.2	6.8	1	6.8
52				None	0
47				None	0
42				None	0
					112.6

TABLE 12.2 (*CONT.*)

September					
Temperature (°F)		h	$\Delta h_{(30-h)}$	Hours	Btu•hr / 1 lbm (air)
Dry	Wet				
77	67	no savings			0
72	65	no savings			0
67	62	28	2	117	234
62	57	24.6	5.4	82	442.8
57	53	22.2	7.8	38	296.4
52	49	19.8	10.2	12	122.4
47	46	18.4	11.6	3	34.8
42				None	0
					1130.4

October					
Temperature (°F)		h	$\Delta h_{(30-h)}$	Hours	Btu•hr / 1 lbm (air)
Dry	Wet				
77	63	28.6	1.4	55	77
72	62	28	2	94	188
67	59	25.8	4.2	131	550.2
62	55	23.2	6.8	141	958.8
57	51	21	9	113	1017
52	47	18.8	11.2	83	929.6
47	42	16.2	13.8	46	634.8
42	38	14.5	15.5	16	248
					4603.4

$$\text{present} = (46{,}675.20 \text{ kwh/year})(1/2.5)(\$.034/\text{kwh})$$
$$= \$634.78/\text{year}$$

$$\text{proposed} = (12{,}080.64 \text{ kwh/year})(1/2.5)(\$.034/\text{kwh})$$
$$= \$164.30/\text{year}$$

$$\text{savings in \$} = \$634.78/\text{year} - \$164.30/\text{year}$$
$$= \$470.48/\text{year}$$

$$\text{savings in kwh} = \frac{46{,}675.20 \text{ kwh/year} - 12{,}080.64 \text{ kwh/year}}{2.5}$$
$$= 13{,}837.82 \text{ kwh/year}$$

$$\text{savings in Btu} = (13{,}837.82 \text{ kwh/year})(3412 \text{ Btu/kwh})$$
$$= 47.22 \times 10^6 \text{ Btu/year}$$

TABLE 12.2 (*CONT.*)

November					
Temperature (°F)		h	$\Delta h_{(30-h)}$	Hours	Btu·hr / 1 lbm (air)
Dry	Wet				
77	60	26.5	3.5	7	24.5
72	60	26.5	3.5	31	108.5
67	57	24.6	5.4	46	248.4
62	53	22.2	7.8	63	491.4
57	49	19.8	10.2	83	846.6
52	45	17.6	12.4	109	1351.6
47	42	16.2	13.8	117	1614.6
42	37	14	16	91	1456
					6141.6

Total	$\dfrac{\text{Btu·h}}{\text{1bm (Air)}}$
March	6,209.5
April	5,472.4
May	1,877.4
June	300.7
July	10.5
August	112.6
September	1,130.4
October	4,603.4
November	6,141.6
	25,858.5 $\quad \dfrac{\text{Btu·h}}{\text{1bm (air)·year}}$

total savings in \$ = savings in kwh + savings in kw + savings in replacement cost
+ savings in heat removal
= (34,594.56 kwh/year)(\$.034/kwh) + (7.56 kw/month)
(12 months/year) × (\$3.99/kw) − \$135.52/year
+ \$470.48/year
= \$1873.15/year

Implementation Cost

Fixtures: (30 fixt.)(\$91.55/fixt.):	\$2746.50
Lights: (30 lights)(\$39.69/light):	\$1190.70
Labor (HPS light addition): (30 fixt.)(1 h/fixt.)(\$15/h):	\$450.00
Labor (incandescent light removal): (68 fixt.)(.5 h/fixt.)(\$15/h):	\$510.00
Total:	\$4897.20

Simple Payback

$$\text{payback} = \frac{\text{cost}}{\text{savings}}$$

$$= \frac{\$4897.20}{\$1873.15/\text{year}} = 2.61 \text{ years}$$

12.3.5 Sawdust Collection Control System

Summary. A wood shop's sawdust collection system is driven by a 20-hp electric motor which runs $8\frac{1}{2}$ h a day. This costs about \$2400 a year for electricity. Also, the system exhausts 6500 ft³ of air per minute, which costs another \$575 annually in increased heating requirements. By installing a control system to shut off the exhaust air to any machine not in use, significant savings will be realized. The recommendation and its savings are presented for consideration.

Recommendations. Install dampers in each of the five vacuum ducts and a speed rheostat at the motor. Used in conjunction with a programmable controller, the damper corresponding to a specific machine will open that duct as the machine is turned on while the speed of the vaccum motor is increased proportionately. One can easily see that when no machines are in use, the motor is completely shut off, providing enormous savings over the present system of 100% load operation. Not only are there savings in electricity but also in the gas used to heat the air that is being evacuated from the plant. The heating plant is assumed to be 70% efficient.

Calculations

Current cost of system:

electricity = demand charge + consumption charge
  = (20 hp)(1 kw/1.34102 hp)(12 months)(\$4.50/kw•month)
    + (20 hp)(1kw/1.34102 hp) $8\frac{1}{2}$h/day)(250 days/year)
    (\$.02915/kwh)
  = \$1729.19

Gas = (6500 ft³/min)($8\frac{1}{2}$ h/day)(60 min/h)(100 heating days/year)
  × (0.7 lb/ft³)(.248 Btu/lb)(65 − 45°F)($\frac{1}{.7}$)(\$3.50/10⁶ Btu)
  = \$5754.84

Cost of ECO:

Speed control	\$1500
Programmable controller	550
Electric dampers (5 at \$65)	325
Wire, switches	50
Installation (estimate)	300
	\$2725

These prices were obtained from local electrical supply stores.

Savings. All concerned machines are operated intermittently, and after discussion with plant personnel, we will assume each machine is used approximately 2 h/day.

gas (heat loss) = (air volume)(operating h reduction)(60 min/h)(heating days/year)

$\times$ (air density)(heat content)(temp. diff.)$(\frac{1}{\text{eff.}})$(gas cost)

= (6500 ft³/min)(8$\frac{1}{2}$ − 2 h/day)(60 min/h)(100 days/year)

$\times$ (.07 lb/ft³)(.248 Btu/lb)(65 − 45°)$(\frac{1}{.7})$($3.50/10^6$ Btu)

= $440/year

electricity = demand savings + consumption savings

= (motor capacity)(kw/1.34102 hp)(12 months/year)

$\times$ (cost/kw·month)(% savings*) + (motor capacity)

$\times$ (kw/1.34102 hp (6$\frac{1}{2}$ h) 250 days/year)(cost/kwh)

= (20 hp)(kw/1.34102 hp)(12 months/year)($4.50/kw·month)

$(\frac{3}{5})$ + (20 hp)(kw/1.34102 hp)(6$\frac{1}{2}$ h)(250 days/year)

(2.915 cents/kwh)

= $1190/year

total annual savings = $440 + $1190

= $1630

Simple Payback

payback = $\dfrac{\text{cost}}{\text{savings}}$

= $\dfrac{\$2725}{\$1630/\text{year}}$ = 1.67 years

12.4 SUMMARY

In this chapter we examined process energy management and how it is similar to/different from space conditioning and lighting energy management. It was shown that process energy management can be much more intricate and complex. The energy manager must understand the entire system and all impacts of any changes.

Next a hierarchy of changes was developed in that the energy manager should try the first on the list, then the second, etc. That hierarchy was eliminate; combine; change person, place, equipment, or sequence; and improve. Examples of each were provided.

---

*The reduction in peak demand is calculated using the assumption that no more than two machines will run continuously and at the same time for 30 min.

Then five abbreviated examples of process modifications for better energy management were given. The reader should not use these examples as typical results but rather as typical types of studies.

BIBLIOGRAPHY

AL-QATTAN, IBRAHIM, "Statistical Analysis of Energy Conservation Audits," unpublished working paper, Oklahoma Industrial Energy Management Program, Stillwater, Okla., 1982.

TURNER, W. C., Senior ed., *Energy Management Handbook,* Wiley-Interscience, New York, 1982.

TURNER, W. C., et al., numerous energy analysis and diagnostic center reports, Oklahoma Industrial Energy Management Program, Stillwater, Okla., 1980–1983.

Chapter 13

Alternative
Energy Sources
and Water Management

13.0 INTRODUCTION

An alternative energy source is one that isn't presently being utilized but that might be usable both technically and economically. Renewable energy sources are those that replenish themselves and so are essentially inexhaustible. For example, coal is an alternative energy source for most areas of the country, but it isn't a renewable source. Solar energy, on the other hand, is both alternative and renewable, as is wind and biomass (refuse). In this chapter we examine alternative energy sources that are also renewable. While they are not a major percentage of energy sources being utilized, their usage is very rapidly growing, and they are generally kind to our environment. The future is likely to see more and more utilization of these sources.

Do they account for much today? Actually, the total utilization is quite small, as depicted later in Table 13–2, but the growth rate is extremely large. If the growth rate depicted in Table 13–2 for solar energy continues over several years, solar energy may indeed become a major source, as may biomass.

In this chapter we examine these potential sources in the following order: solar-active, solar-passive, solar-photovoltaic, wind, and refuse. Emphasis is on applications in the industrial-commercial environment.

In the last part of the chapter we discuss management of another vital and renewable resource: water. Water will be a major crisis someday soon. The manager skilled in water management will be prepared to meet this challenge.

TABLE 13-1 AVERAGE SOLAR RADIATION FOR SELECTED CITIES

City	Slope	Average daily radiation (Btu/day·ft²)											
		Jan.	Feb.	March	Apr.	May	June	July	Aug.	Sept.	Oct.	Nov.	Dec.
Albuquerque, NM	Hor.	1134	1436	1885	2319	2533	2721	2540	2342	2084	1646	1244	1034
	30	1872	2041	2295	2411	2346	2390	2289	2318	2387	2251	1994	1780
	40	2027	2144	2319	2325	2181	2182	2109	2194	2369	2341	2146	1942
	50	2127	2190	2283	2183	1972	1932	1889	2028	2291	2369	2240	2052
	Vert.	1950	1815	1599	1182	868	754	795	1011	1455	1878	2011	1927
Atlanta, GA	Hor.	839	1045	1388	1782	1970	2040	1981	1848	1517	1288	975	740
	30	1232	1359	1594	1805	1814	1801	1782	1795	1656	1638	1415	1113
	40	1308	1403	1591	1732	1689	1653	1647	1701	1627	1679	1496	1188
	50	1351	1413	1551	1622	1532	1478	1482	1571	1562	1679	1540	1233
	Vert.	1189	1130	1068	899	725	659	680	811	990	1292	1332	1107
Boston, MA	Hor.	511	729	1078	1340	1738	1837	1826	1565	1255	876	533	438
	30	830	1021	1313	1414	1677	1701	1722	1593	1449	1184	818	736
	40	900	1074	1333	1379	1592	1595	1623	1536	1450	1234	878	803
	50	947	1101	1322	1316	1477	1461	1494	1448	1417	1254	916	850
	Vert.	895	950	996	831	810	759	791	857	993	1044	842	820
Chicago, IL	Hor.	353	541	836	1220	1563	1688	1743	1485	1153	763	442	280
	30	492	693	970	1273	1502	1561	1639	1503	1311	990	626	384
	40	519	716	975	1239	1425	1563	1544	1447	1307	1024	662	403
	50	535	723	959	1180	1322	1341	1421	1363	1274	1034	682	415
	Vert.	479	602	712	746	734	707	754	806	887	846	610	373

TABLE 13.1 (CONT.)

City	Slope	Average daily radiation (Btu/day•ft²)											
		Jan.	Feb.	March	Apr.	May	June	July	Aug.	Sept.	Oct.	Nov.	Dec.
Ft. Worth, TX	Hor.	927	1182	1565	1078	2065	2364	2253	2165	1841	1450	1097	898
	30	1368	1550	1807	1065	1891	2060	2007	2097	2029	1859	1604	1388
	40	1452	1601	1803	1020	1755	1878	1845	1979	1995	1907	1698	1488
	50	1500	1614	1758	957	1586	1663	1648	1820	1914	1908	1749	1549
	Vert.	1315	1286	1196	569	728	679	705	890	1185	1459	1509	1396
Lincoln, NB	Hor.	629	950	1340	1752	2121	2286	2268	2054	1808	1329	865	629
	30	958	1304	1605	1829	2004	2063	2088	2060	2092	1818	1351	1027
	40	1026	1363	1620	1774	1882	1909	1944	1971	2087	1894	1450	1113
	50	1068	1389	1597	1679	1724	1720	1763	1838	2030	1922	1512	1170
	Vert.	972	1162	1156	989	856	788	828	992	1350	1561	1371	1100
Los Angeles, CA	Hor.	946	1266	1690	1907	2121	2272	2389	2168	1855	1355	1078	905
	30	1434	1709	1990	1940	1952	1997	2138	2115	2066	1741	1605	1439
	40	1530	1776	1996	1862	1816	1828	1966	2002	2037	1788	1706	1550
	50	1587	1799	1953	1744	1644	1628	1758	1845	1959	1791	1762	1620
	Vert.	1411	1455	1344	958	760	692	744	918	1230	1383	1537	1479
New Orleans, LA	Hor.	788	954	1235	1518	1655	1633	1537	1533	1411	1316	1024	729
	30	1061	1162	1356	1495	1499	1428	1369	1456	1490	1604	1402	1009
	40	1106	1182	1339	1424	1389	1309	1263	1371	1451	1626	1464	1058
	50	1125	1174	1292	1324	1256	1170	1137	1259	1381	1610	1490	1082
	Vert.	944	899	847	719	599	546	548	647	843	1189	1240	929
Portland, OR	Hor.	578	872	1321	1495	1889	1992	2065	1774	1410	1005	578	508
	30	1015	1308	1684	1602	1836	1853	1959	1830	1670	1427	941	941
	40	1114	1393	1727	1569	1746	1739	1848	1771	1680	1502	1020	1042
	50	1184	1442	1727	1502	1622	1594	1702	1673	1651	1539	1073	1116
	Vert.	1149	1279	1326	953	889	824	890	989	1172	1309	1010	1109

Source: Reproduced from Reference 2.

13.1 SOLAR-ACTIVE

13.1.1 Amount

Approximately 430 Btu/h•ft² of solar energy hits the earth's atmosphere. Because of diffusion in the atmosphere and clouds, this is greatly reduced to somewhere around a maximum of 300 Btu/h on the earth's surface at 40°N latitude. This maximum, of course, only occurs at certain times of the day and year, so the average is significantly less. However, at this rate a set of collectors designed to develop 1×10^6 Btu/h of energy would have to be 3333 ft² in size without allowing for cloudiness, variances throughout the day, or collector efficiency. If the collector were tracking the sun throughout the day, it might be able to gather 8×10^6 Btu/day. Assuming it operates 365 days/year, the collector would be able to harvest an absolute maximum of

$$(8 \times 10^6 \text{ Btu/day}) (365 \text{ days/year}) = 2920 \times 10^6 \text{ Btu/year}$$

At $4.00/10^6$ Btu, this energy would be worth about $11,700/year. The necessary collector space is 3333 ft², and the installed cost including controls might be around $20/ft². Therefore, the cost of the proposed collectors would be around $66,000, making the payback *under ideal conditions* somewhere around 5.6 years. Actual conditions would likely require a significantly larger collector, as will be shown below.

Detailed calculations of the amount of solar energy striking a surface located at a given latitude and tilted at a certain angle require knowledge of several angles, including the solar altitude angle, the solar azimuth angle, and the tilt angle. The values all vary with time of day, month, location, tilt of collector, etc., so the detailed calculations are tedious but quite simple. To save time, several authors [1,2] have prepared tables to help a person determine the amount of solar energy available. Table 13–1 is an example of such a table.

Considering Table 13–1 for Lincoln, Nebraska (approximately 40° N latitude), we find the solar radiation on a surface tilted at 40° to be about $.6 \times 10^6$ Btu/ft²•year. Assuming a collector is 70% efficient, the energy available is about $.42 \times 10^6$ Btu/ft²•year. Thus the 3333-ft² collector used would supply about 1400×10^6 Btu/year. At $4.00/10^6$ Btu, the savings would be $5600/year, lengthening the payback to

$$\frac{66,000}{5600} = 11.8 \text{ years}$$

under expected conditions.

Various tax incentives are available at both the federal and local levels to speed up this payback, but solar energy still has a way to go before its economics are universally attractive. Solar energy is coming, however. Consider Table 13–2, drawn from data developed by the BioEnergy Council. Solar energy is growing rapidly, but note that its overall level is still quite low.

TABLE 13–2 ENERGY SOURCE CHANGE, 1975–1980

	10^{15} Btu, 1975	10^{15} Btu, 1980	% change
Hydroelectric	3.164	2.915	−7.9
Nuclear	1.900	2.704	+42.3
BioEnergy	1.7003	2.484	+45.9
Solar & wind	.003	.014	+367.0
All other	65.584	70.558	+7.6
All energy	72.354	78.675	+8.7

13.1.2 Collectors

A solar collector is a device used to thermally collect, store, and move solar thermal energy. A wide variety of different types exists, as demonstrated in Figure 13–1. In Figure 13–1, tracking types of solar collectors can be broken down in much more detail, but almost all industrial uses to date have been flat plate, so this discussion will concentrate there.

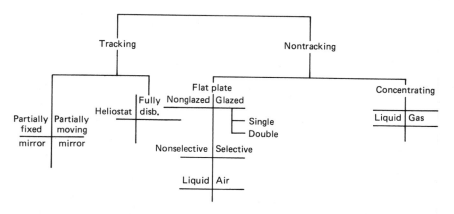

Figure 13-1 Types of solar collectors.

A flat-plate collector is a device almost always facing to the south (northern hemisphere) and tilted at some angle. Its purpose is to allow the sun's shortwave radiation to enter the collector and heat some fluid [usually water (with or without antifreeze) or air]. The hot fluid is then pumped from the collector to the point of use or to storage for later use.

A typical flat-plate solar collector is demonstrated in Figure 13–2. A typical application is given in Figure 13–3. In this application, solar energy heats an ethylene-glycol mixture that is pumped to a storage tank. The tank then heats water through a heat exchanger for alternative use as shown. In some applications, such as preheating boiler makeup water, the water itself can be pumped through the collector to a stor-

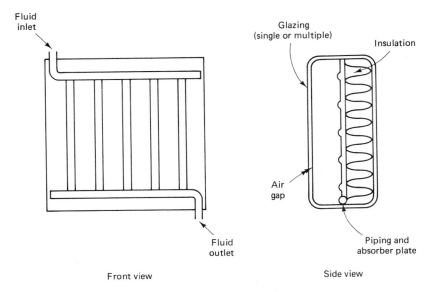

Figure 13-2   Flat-plate solar collector.

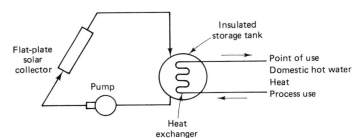

Figure 13-3   Typical flat-plate solar collector application.

age tank or directly to the boiler room. In such applications, care must be taken to prevent freezing (drain down provisions are usually employed).

The collector may be glazed or nonglazed. The glazing (usually tempered glass) is designed to allow the shortwave radiation of the sun to enter the collector but prohibit the longer-waved reradiation waves from leaving. This heats up the space (greenhouse effect) and increases the efficiency of the collector but, unfortunately, also the cost. Dual glazing would further cut down the heat loss while not appreciably restricting incoming solar energy. Unglazed collectors are quite efficient at lower-temperature applications (such as swimming pool heaters), while glazed collectors are more efficient at higher temperatures. Two glazing layers help efficiency at even higher temperatures. Typical efficiencies are given in Figure 13-4.

The selectivity of the collector surface is an important property. The collector must be able to absorb shortwave radiation readily and emit long-wave radiation stingily. Surfaces that have a high shortwave absorption and a low long-wave emit-

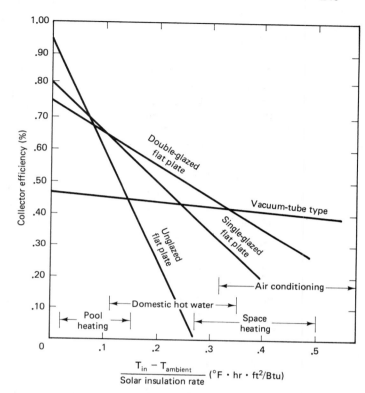

Figure 13-4 Typical glazing efficiencies.

tance are called *selective surfaces*. As might be expected, selective surfaces perform better at higher temperatures than do nonselective surfaces. In fact, a single glazed selective surface collector has efficiencies very similar to a double-glazed nonselective surface collector.

The medium chosen to cool the collector or, more accurately, to move the thermal energy from the collector to the point of use or to storage can be a liquid or air. Thus far, the examples presented have assumed the fluid is a liquid, but many applications are suitable for air. For example, an air solar collector could be used as an air preheater (recuperator) for an industrial furnace or boiler. To the authors' knowledge, this application is not widespread, but it does have the interesting characteristic of matching solar time and time of use (i.e., when the sun is shining, the energy is most likely needed).

It is much more difficult and expensive to move thermal energy with air than with liquid. In fact, the horsepower required to move the same amount of thermal energy may be 10 times higher for an air system than for a liquid system.

Air does have the advantage that it won't freeze. In liquid systems, ethylene-glycol or some other antifreeze must be used, or the system must have well-designed drain down controls. Also, air systems do not have corrosion problems, and leaks do not present as much of a problem as with liquid systems. Air systems do have lower heat transfer rates, so the system must be carefully designed to provide a sufficient heat transfer surface.

A need for even higher heat (temperatures of 250°F or higher) usually requires a concentrating collector. A concentrating collector is usually a tracking collector due to the need to keep the sun's rays concentrated on a small surface. A typical design for a parabolic trough-type collector is shown in Figure 13-5. The collector can track in an east to west direction to follow the daily sun, in a north to south direction to follow the seasons, or both. Obviously, the better tracking units will be more efficient.

The surface of a concentrating collector must be highly reflective, enabling concentration of the sun's rays on the heat absorption device. The heat transfer fluid can be a liquid or gas.

Other types of tracking collectors use movable mirrors that can concentrate the solar energy on a small surface that remains fixed. The *power tower* is such an application where the receiver surface is a tower. Tracking mirrors are located on the ground around the base of the tower. These fully tracking mirrors are usually computer-controlled to concentrate the maximum amount of solar energy on the tower. Applications are mainly for steam generation, usually electric power. This type of application takes a large amount of land area and requires careful maintenance. Wind damage is also a problem. As might be expected, there are very few power towers in operation. There is one in New Mexico for experimentation purposes.

13.1.3 Storage

One of the biggest obstacles to widespread solar utilization is the fact that timing very seldom matches. That is, the solar energy is available when it's not needed, and it's not available when it is needed, for example, maximum heat is usually needed when the sun is not shining, especially at night. Also, solar energy flows cannot be regulated. When the sun shines, the collector delivers energy.

For these reasons, solar applications usually require some type of storage sys-

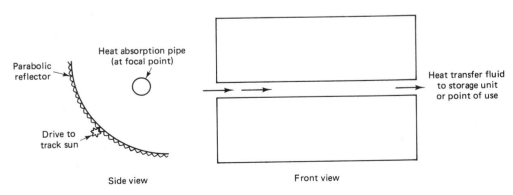

Figure 13-5 Parabolic trough solar collector.

tem. The storage system becomes both a storage for later use and a regulator to control energy flows. Figure 13-3 depicted one possible liquid storage system.

There are three basic types of storage systems:

- *Liquid.* Liquid storage systems normally utilize water or a water-antifreeze mixture. The storage capability is determined by the sensible heat of the liquid (for water it would be 1 Btu/lb•°F).
- *Rocks.* Used for air systems, rock storage uses the sensible heat content of rocks for storage. Typically, airflow is top to bottom for storage and bottom to top for use as needed.
- *Phase change materials.* Both the preceding systems utilized sensible heat. This system utilizes the larger latent heat in phase changes (e.g., the melting of ice). The required storage volume is smaller, but the cost is higher. Eutectic salts are often used.

13.1.4 Industrial Applications

There simply has not yet been a large-scale incorporation of solar energy in industry. There are several projects throughout the country, but most are on a pilot-experimental basis. Very few are justifiable economically, but each day brings the time of cost-effective applications a bit closer.

Following are some applications of solar thermal energy in industry. A discussion is given of their relative cost effectiveness.

- *Solar-heated asphalt storage tanks.* Previously, a company had used portable propane burners to keep the asphalt hot. Solar collectors and thermal insulation keep the asphalt at the required temperature. Savings and convenience made this application almost cost effective.
- *Solar-heated hot water.* Undoubtedly the bright spot today is solar-heated or -augmented hot water. The hot water tank itself becomes the storage (or at least a part of it), and the hot water is needed the entire year. In some parts of the country, solar water heaters are very cost effective—especially where ample tax credits are available. Solar-augmented process feeds would be similar to this idea, as would be solar-heated makeup water for boilers and cleaning tanks.
- *Solar recuperators.* Industrial furnaces also require heat year-round, so recuperation (combustion air preheating) is a likely candidate. In installations where outside air might already be used, the application is very simple and probably requires no storage. When the sun shines, the air is preheated; otherwise it is not. (This may cause problems with burners and excess air control.) The authors worked with one company that designed such a system. In an area of the country where gas costs were small and few tax incentives existed, the payback was around 10 years. This idea offers some promise for the medium-term future.

- *Solar space heat.* Most of the applications to date involve solar space heat with its problems of lack of year-round need. Substantial tax credits (often 60–70%) make this feasible in areas where the credits exist, but the higher credits are usually for residential use, not industrial use. In the absence of tax credits, active solar collectors for space heating are very difficult to justify.
- *Solar air conditioning.* Some places have attempted to remedy the short time of need problem by using higher-temperature collectors and air conditioning through an absorption process. To the authors' knowledge, these uses are not cost effective at this time.

13.2 SOLAR-PASSIVE

If the discussion on active solar systems disappointed you, then this section should perk you back up. Passive solar systems are often cost effective today without the benefits of tax credits. The future is likely to see many more of them. Passive solar energy is the use of solar energy in a direct fashion rather than requiring active solar collectors, pumps, separate storage facilities, etc. For example, a south-facing glass window is the solar collector and the interior of the building the storage device. As will be shown in this section, careful attention to solar energy in building design can reduce energy costs significantly.

To examine the potential impact, consider a light manufacturing building facing south. The building has a total wall area (minus glass) of 2000 ft² and a roof area of 3000 ft². The R values of the roof and walls are 22 and 12, respectively. There is 300 ft² (about 40% of the south wall) of south-facing double-pane overhung glass so as to permit full sunlight for the 5-month heating season. Assume the plant is in Fort Worth, Texas. The glass transmits 80% of the solar energy hitting it.

From Table 13–1 (or other sources) we find the total solar load on a vertical surface for the months of November, December, January, February, and March to be

$$(80\%)\,(300\text{ ft}^2)\left[1509(30) + 1396(31) + 1315(31) + 1286(28)\right.$$
$$\left. + 1196(31)\frac{\text{Btu}}{\text{ft}^2\text{•year}}\right]$$
$$= \underline{48.58 \times 10^6 \text{ Btu/year}}$$

The heat loss (HL) (assuming no setback) is

$$\text{HL walls} = \left(\frac{1\text{ Btu}}{12\text{ h•}^\circ\text{F•ft}^2}\right)(2000\text{ ft}^2)\left(\frac{2363^\circ\text{F days}}{\text{year}}\right)\left(\frac{24\text{ h}}{\text{day}}\right)$$
$$= \underline{9.45 \times 10^6 \text{ Btu/year}}$$

$$\text{HL roof} = \left(\frac{1\text{ Btu}}{18\text{ h•}^\circ\text{F•ft}^2}\right)(3000\text{ ft}^2)\left(\frac{2363^\circ\text{F days}}{\text{year}}\right)\left(\frac{24\text{ h}}{\text{day}}\right)$$
$$= \underline{9.45 \times 10^6 \text{ Btu/year}}$$

$$\text{HL glass} = \left(\frac{1 \text{ Btu}}{1.85 \text{ h} \cdot {}^{\circ}\text{F} \cdot \text{ft}^2} \right) \ (300 \text{ ft}^2) \ \left(\frac{2363 {}^{\circ}\text{F days}}{\text{year}} \right) \left(\frac{24 \text{ h}}{\text{day}} \right)$$

$$= \underline{9.20 \times 10^6 \text{ Btu/year}}$$

$$\text{total HL} = \underline{28.1 \times 10^6 \text{ Btu/year}}$$

Placing glass on 40% of the south-facing wall then can supply 170% of the total heat loss for the year. Now practicality enters:

- The heat loss by infiltration may be as large as the total heat loss through the roof and walls—especially if there is much exhaust air.
- The sun doesn't shine some days, so the heating plant will have to be designed as large as it would be with no solar aid.
- On bright sunny days, the building might get too hot, so the glass area might have to be reduced or adjustable shading used.
- This building is fairly well insulated (though not excessively). Many manufacturing buildings have little insulation.
- All glass is on the south wall. There may be a need to place some on the other walls.
- Fort Worth may not be typical of the rest of the country.

The point is made, however, that passive solar energy can contribute a large percentage of the heating required. More attention needs to be paid to passive solar utilization in the design of buildings. This is especially true for manufacturing buildings whose hours of operation normally coincide well with sun hours. At night, the thermostat can be substantially reduced, whereas in homes, night setback can't be used as readily.

Following is a list of ways the industrial manager might use passive solar energy (or avoid unwanted solar loads). (We have also included some ideas that save in ways other than utilizing solar energy.)

1. *Locate energy-efficient property.* The facility should be located near energy supplies (to minimize transmission losses as well as costs) and *near good transportation facilities* and *not in energy-intensive spots* (avoid too windy areas, areas on tops of hills, areas on north slopes). However, these spots may become attractive in the future as renewable energy sources.

2. *Locate facility properly on the property.* Again, the facility should be near major transportation facilities. Advantage should be taken of unique spots that may offer shading in the summer sun and/or windbreaks in the winter through the use of existing deciduous trees or other natural properties. Hills can be utilized as berms to improve insulation instead of spending money to level.

3. *Landscape to conserve energy.* Much energy in HVAC can be saved through proper landscaping. *Avoid asphalt or concrete areas around the building* as

much as possible, since grasses, shrubs, and vines are much cooler in the summer. Place deciduous trees strategically, as they offer shade during the summer, yet allow sunlight to penetrate in the winter. *Use vines and shrubs to offer additional shading.* In fact, thick shrubs placed close to a building increase the *R* value of the walls. *Use trees and shrubs as windbreaks and wind diverters.* For example, evergreen trees or shrubs off the northwest corner of the building can break the cold winter winds and divert the summer breezes for better utilization, as shown in Figure 13-6.

4. *Consider the use of underground structures. Use a large amount of backfill* on northern and western walls. *Partially submerge* the entire structure, or possibly *submerge the entire building* as with the warehouses in the caverns under Kansas City.

5. *Orient the facility for energy conservation.* The building should face south. The shorter dimension should run north to south and the longer, east to west. The manufacturing plant in Figure 13-6 demonstrates this. This allows minimum sun exposure in the summer on the east and west walls. However, during the winter, because of the lower sun angle, the sun helps to heat the facility. With proper placement of windows, natural ventilation is also feasible.

6. *Insulate well. Install insulation* in the ceiling and the walls as well as on slab (if appropriate). Check for local recommended levels. Figure 13-7 illustrates recommended insulation placement for an underground manufacturing facility.

7. *Minimize wall perimeter area.* Use regular-shaped buildings—square or rectangular—to minimize the wall area and heat loss.

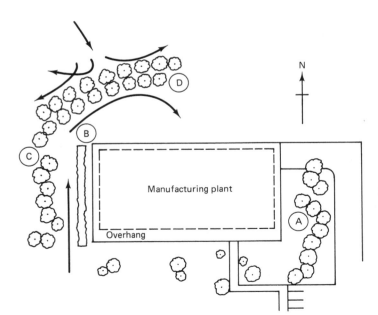

Figure 13-6 Landscaping for energy conservation. (a) Deciduous trees: In summer, allow the early morning sun to penetrate and then protect; in winter, allow the sun to penetrate. (b) Hedge: Catch the late afternoon rays. (c) Deciduous trees: Provide shade in the afternoon. (d) Shrubs or tall hedge: In winter, block northwest winds; in summer, help divert southwest winds around the building.

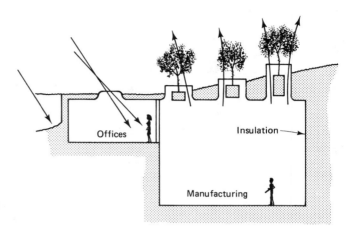

Figure 13-7   Recommended insulation placement.

8. *Consider the installation of insulating glass or storm windows.* This is especially true in dense personnel areas such as offices.

9. *Use minimum amount of glass—well placed.* Avoid glass on northern and western sides. Southern exposure with proper solar influx planning is best (as discussed earlier).

10. *Consider* the use of windows that can be opened to utilize natural ventilation.

11. *Consider* the use of tinted window glass but on walls other than south—especially east and west.

12. *Utilize overhangs or awnings.* On southern exposures, overhangs block summer sun but allow winter sun to enter as shown in Figure 13-8. Passive solar systems as shown help to further reduce energy costs, and they provide attractive warm areas for personnel. Any good architect can tell you the proper amount of overhang to allow for sunlight in winter months. It varies with location.

13. *Design roof carefully.* Use light colors in warm climates and dark in cold. De-

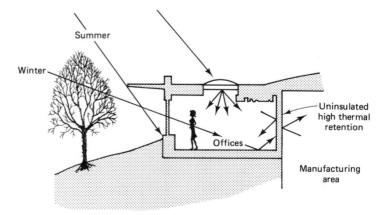

Figure 13-8   Energy conservation through the use of overhangs.

sign so it can be sprayed in the summer. Be careful of flooding roofs—they will leak eventually.

14. *Engineer wall openings. Minimize the number of openings. Caulk and weather-strip doors and windows well. Utilize insulating pads on dock doors.* Consider using adjustable dock pads. Consider interlocking doors and heating units so that when doors are open, the heat is off. Utilize automatic doors or various types of "see through" materials (plastic strips, plexiglass, etc.) for doors that must be used frequently. Use *good pedestrian doors to avoid using large dock doors for pedestrian traffic. Utilize revolving doors or entrance vestibules* to minimize air infiltration. Design overhead doors with two-position openings to match trucks and/or use adjustable dock seals.

15. *Locate facilities within the plant to minimize energy required for personnel comfort.* High personnel density departments should probably be located in southern exposure areas of the plant and not in northern or western areas. Avoid northern or western exposure for dock areas (shipping and receiving). Figure 13-9 demonstrates one possible layout considering energy requirements only. Plan the layout so exhaust air from one area can be used in another (e.g., maybe hot air at ceiling height in one area can be used as combustion air for a large furnace). Finally, arrange facilities so that energy control is facilitated (e.g., lights and motors can be switched off in one location).

16. *Engineer waste heat recovery systems into the facility design.* Waste heat is the one renewable energy source today whose utilization is frequently cost effective. *It is much easier to incorporate recuperators or feedstock preheating apparatus* in the initial design rather than retrofitting. Boilers, furnaces, large motors being cooled, lighting fixtures, any cooling fluids, and compressors are

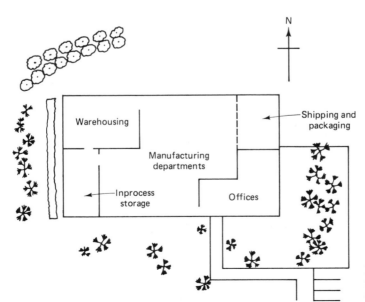

Figure 13-9 Hypothetical plant layout to minimize energy requirements for space conditioning.

just a few of the potential sources for waste heat recovery. By considering waste heat recovery in the design phase, sources of waste heat can be located such that heat recovery is facilitated. This can be done through location of waste heat producing equipment where it can be utilized and such that heat recovery equipment can be installed.

17. *Consider departmental or cost center metering of utilities.* By doing this, each cost center can be held accountable for its energy consumption, and energy can become a part of the budgeting process. This requires extensive preconstruction planning to be able to lay lines and install meters as needed.

18. *Locate all boilers, hot water tanks, and other heated tanks to minimize distribution distances and consequently energy loss in distribution.*

19. Design the boiler location and steam distribution system to facilitate *return of condensate* and/or reflashing or low-pressure steam.

20. *Utilize drapes or outside partitions to further insulate windows* and to reduce solar load when desired. Several examples of outside partitions are shown in Figure 13-10.

21. Place air compressors for easy maintenance, to use fresh cold air, and to minimize transmission losses. Utilize *step-up* air compressors to be able to reduce plant-wide pressures.

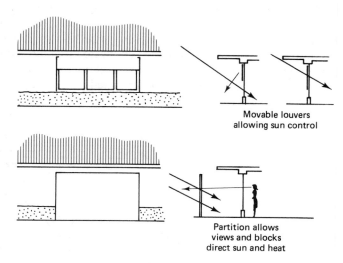

Movable louvers
allowing sun control

Partition allows
views and blocks
direct sun and heat

Figure 13-10   Outside partitions.

13.3 SOLAR-PHOTOVOLTAICS

Photovoltaics is the conversion of sunlight to dc electricity through a photocell. Historically, photovoltaics has not been cost effective, but the needs for photovoltaics in space and the subsequent research coupled with rising traditional energy costs have pushed photovoltaics ahead. Although still not normally cost effective, photovoltaics is closer.

Generally there are three types (sizes) of systems:

1. *Small* (1–10 kw). These photovoltaics would be suitable for remote locations or other locations where conventional electricity may be costly. Examples are irrigation pumps, construction sites, communications equipment, and field battery charging.
2. *Medium* (10–1000 kw). These photovoltaics could be utilized by larger industrial facilities and/or remote communities and could also be used as supplemental or peaking power.
3. *Large* (megawatts). These photovoltaics could be utility-owned large-scale power-generating stations probably located in desert areas.

Industry seems to offer applications for photovoltaics, but the usage is limited in size and scope. Ideal applications would have the following characteristics:

1. Equipment operates on dc power.
2. Substantial sunlight is available.
3. Need can tolerate random losses of power.
4. Duty cycles match (i.e., power is needed when the sun is shining).

Photovoltaic cells are semiconductor devices which can convert the energy in photons of light into dc electrical energy. Most cells are made from single-crystal high-purity silicon and small amounts of trace elements such as boron and phosphorus. These elements are combined in such a way as to create a material with excess electrons (*n*-type semiconductor) connected to a material with insufficient electrons (*p*-type region). The area where the materials are joined together can be linked to a battery. As long as the area is illuminated by light, electrical energy will be produced (see Figure 13–11).

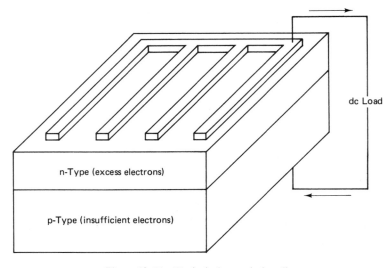

dc Load

n-Type (excess electrons)

p-Type (insufficient electrons)

Figure 13–11   Typical photovoltaic cell.

The electrical energy generated by most cells is about .5 v and a current that varies with the area of the cell and amount of light. Typical wattage is about 12.5 w/ft² with a conversion efficiency of 7–15%.

To achieve necessary voltages and currents, solar cells are combined in series and parallel, like batteries. Cells are usually placed in series to meet the necessary voltage, and then multiple strings are placed in parallel to develop the desired current (see Figure 13–12). Because of variance in time of need and solar intensity, some type of storage device and voltage regulation is necessary. Normally, a chemical battery storage system fills this need admirably. Finally, a backup system is often needed to allow for consecutive cloudy days.

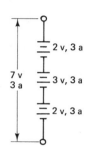
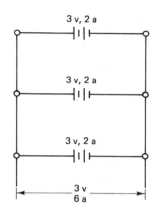

Figure 13–12   Photovoltaic hookups.

13.4 WIND ENERGY

Much has been said about wind energy and its potential. One reference [3] states that the potential in the United States for wind energy is many times greater than the present consumption of electrical energy, but, of course, this potential will never be obtained due to aesthetic, construction cost and radio-TV interference problems.

Almost nothing has been done to utilize wind in industry. Usually industry is located in congested areas, so windmills would not be popular. Also, industry is not likely to develop wind energy as a source when it is simply not presently cost effective.

The power density of wind is given by

$$\frac{P}{A} = \frac{1}{2}\rho V^3$$

where A = area normal to the wind
ρ = density of air (about .076 lb/ft³)
V = velocity of air
P = power contained in the wind

This can be rewritten as follows (K is a constant for correcting units):

$$\frac{P}{A} = KV^3$$

where $K = 5.08 \times 10^{-3}$

$$\frac{P}{A} = \text{w/ft}^2$$

$$V = \text{mph}$$

Unfortunately, only a small percentage of this power can be obtained. It can be shown that theoretically .5926 of the power can be extracted, but practically only 70% or so of that can be obtained. Consequently, only .70(.5926) or about 40% of power is harvestable.

For example, a 20-mph wind could yield about

$$\frac{P}{A} = (5.08 \times 10^{-3})(40^3)(.40) = 130 \, \text{w/ft}^2$$

These figures demonstrate why it is necessary to find areas with consistent high wind velocities and to use large windmills or a series of smaller ones.

Wind speed offers other problems. Too little wind won't initiate power output for most windmills. (The wind speed must be above a "cut-in" speed unique to the windmill.) Too much wind creates other problems that could conceivably destroy the windmill, so most windmills feather out at some high wind speed.

13.5 REFUSE-DERIVED FUEL

Probably the most hopeful area for renewable fuel utilization in industry is refuse-derived fuel. The concept is not new, but widespread adoption is just now beginning. About 70% of the typical household refuse is combustible, but industrial wastes are usually more combustible. In fact, some industrial waste is a fairly high-quality fuel, as pulp and paper mills have been proving for years.

Combusting waste also reduces the volume of the waste by 80% or more and often makes it easier to handle. Consequently, one of the big savings for refuse-derived fuel is in reduced disposal costs. In some industries, this savings may be much larger than the actual Btu content.

Some typical heating values for various types of industrial wastes are shown in Figures 13–13 and 13–14. As can be seen in these tables, many sources of industrial waste have significant Btu content. Refuse-derived fuel is certainly an interesting fuel source worthy of study.

Refuse-derived fuel from municipal waste is attracting a great deal of interest today. Some successful applications exist today, as do some failures. In this book we are concentrating on industrial-institutional energy management, so this subject will be left to other books. In the rest of this section we will concentrate on industrial-institutional applications.

	Wood, average, seasoned (%)	Wood waste, Douglas fir (%)	Hogged fuel, Douglas fir (%)	Sawdust (%)
Proximate analysis:				
Moisture	24.0	35.9	47.2	44.9
Volatile matter	65.5	52.5	42.9	44.9
Fixed carbon	9.5	11.1	8.9	9.5
Ash	1.0	.5	1.0	.7
Ultimate analysis:				
Hydrogen	7.2	8.0		
Carbon	37.9	33.5		
Nitrogen	.1	.1		
Oxygen	53.8	57.9		
Sulfur	—	0		
Ash	1.0	.5		
High heating value (Btu/lb)	6300	5800	4670	4910

Figure 13–13 Wood waste characteristics. (From *Instructions For Energy Auditors,* U.S. Dept. of Energy, 1978.)

	Heating value	
Fuel	Btu/lb	Btu/ft³
Solid fuels:		
Bagasse	3,600– 6,500	
Bark	4,500– 5,200	
Wood waste	4,500– 6,500	
Sawdust	4,500– 7,500	
Coffee grounds	4,900– 6,500	
Rice hulls	5,200– 6,500	
Corn cobs	8,000– 8,300	
Municipal refuse	4,500– 6,500	
Industrial refuse	6,600– 7,300	
Coal	8,000–24,000	
Liquid fuels:		
Black liquor (pulp mills)	4,000	
Dirty solvents	10,000–16,000	
Gasoline	20,700	
Industrial sludge	3,700– 4,200	
Naphtha	20,250	
Naphthalene	18,500	
Oil waste	18,000	
Paints and resins	6,000–10,000	
Spent lubricant	10,000–14,000	
Sulfite liquor	4,200	
Oil	18,500	
Gas fuels:		
Coke, oven gas		500–1000
Refinery gas		1200–1800
Natural gas		1000

Figure 13–14 Heating values of industrial waste fuels. (From *Instructions For Energy Auditors,* U.S. Dept. of Energy, 1978.)

The following is offered as a procedure to be used in analyzing waste fuel sources:

Step 1. Determine the heating value and quantity of the waste.

Step 2. Determine the technical feasibility of utilizing the waste (burning, pyrolysis, anaerobic digestion, etc.). Include necessary pollution control and ash disposal.

Step 3. Develop a system design including waste fuel handling, preparation, firing, and disposal. Estimate the cost involved.

Step 4. Do economic analysis including all incremental costs (step 3) and all savings (over conventional fuel). Don't forget savings in disposal cost and any tax incentives (tax credits or preferred depreciation schemes) from federal and state governments.

After going through this procedure several times (at least once for each alternative), the proper decision should emerge.

At some stage, the analysis should include the incorporation of other industry's or even municipal waste. This way economics of scale might be obtained. However, extreme care should be taken to ensure that these sources are reliable in quantity, flow rate, and quality.

The technology for utilizing waste fuel is rapidly becoming more developed in that equipment is commercially available. Some of the technologies receiving attention include the following:

- Burning raw refuse in water well incinerators
- Shredding, pelletizing, or otherwise preparing fuel and burning in boilers
- Anaerobic digestion (conversion of waste to gas in an oxygen-free atmosphere)
- Pyrolysis (thermal decomposition) of waste in the absence of oxygen

In closing this section, summaries of some recent applications of refuse-derived fuel are given:

- The state of Texas (Department of Corrections) is installing incinerator-generator systems at two of its state prisons. The systems will use commercially available equipment to burn the unprepared refuse. Each unit will produce up to 8000 lb/h of 225-psig steam by burning about 2600 lb of refuse per hour. Refuse will be brought in from other prisons. Payback is estimated at 3 years.
- A rice milling company in Louisiana is installing a boiler-steam turbine system to burn about 250 tons of hulls per day, producing 115,000 lb of steam per hour and about 11.5 Mw of electricity. The electricity will be sold to a local utility, generating about 3.3×10^6 in revenue per year. Although no use presently exists for the low-pressure steam emerging from the turbine, it is available. Payback is estimated at 4 years [4].

- A tire retread company in the northeast is installing equipment to turn its waste tires into oil and gas through pyrolysis. The system will eliminate the problem of tire disposal (landfills don't want tires as they don't decompose and tend to work their way back to the surface over the years) and generate enough oil and gas to fire its boilers. After considering fuel and landfill savings as well as 20% tax credits from the federal government, payback is estimated at 5 years [5].

13.6 WATER MANAGEMENT

Water management is the efficient and effective use of water to accomplish objectives. Like energy management, the main goal is to improve profits or reduce costs. It is included in this text on energy management because water and energy utilization are intricately entwined in most organizations. As water consumption goes up, so does energy, and vice versa. Water management is included in this chapter on renewable energy sources as water is renewable and is often an efficient energy source (water in cooling towers is a very inexpensive source of cooling).

Why worry about water? The sheer volume of water used in this country each day is staggering. Each person in the United States consumes directly and indirectly almost 90 gal of water per day. New York City uses about 1.5×10^9 gal/day. At the same time, water costs are going up rapidly, and some sources are drying up (e.g., Midwestern aquifers).

Yet industry can save dramatic amounts of money and water through attention to its water-consuming equipment. In this section we will demonstrate how to save dollars and gallons of water in industry.

There are three primary types of water users in industry. They are boilers, cooling towers, and process equipment.

In boilers the primary ways to save are the following:

- *Blowdown.* Blowdown should be reduced to the minimum possible, which is determined by the feedwater quality and the amount of condensate return. This also saves dramatic amounts of energy.
- *Condensate return.* Condensate should be returned to the boiler whenever economically feasible. This saves water by reducing the amount of makeup and by reducing the amount of blowdown required. Returning condensate also saves significant amounts of energy and water treatment costs.
- *Steam leaks and steam traps.* As already demonstrated, steam leaks are extremely expensive in energy losses, but significant amounts of water are also lost in steam leaks. (An estimate of the amount can be obtained by dividing the annual loss in Btus by the enthalpy of evaporation.) Stuck-open steam traps are steam leaks if the condensate is not returned and are still wasteful even if it is returned.

In cooling towers, the primary ways to save are the following:

- *Bleed.* Bleed in a cooling tower is almost identical to blowdown in a boiler. The purpose is to prevent impurity buildup. Bleed should be reduced to a minimum and reused if possible. (Sometimes bleed can be used to water lawns or as rinse water makeup. Careful attention must be paid to the chemistry.)
- *Sewerage charges.* Often sewerage charges are based on the amount of water consumed. Yet water consumed in a cooling tower does not go into the sewer. Negotiations with municipalities can often reduce the sewerage charge dramatically if large cooling towers are present. (Usually about 1% of the flow rate must evaporate for each 10°F drop in water temperature.)
- *Preventive maintenance.* As baffles become broken or clogged with dirt and slime, cooling capability drops dramatically. Since a 1°F drop in condensing water temperature can mean a 3% saving in electrical chiller input, preventive maintenance of towers is important.
- *Tower water.* Often cooling tower water can be used directly for process cooling instead of using chilled water. Whenever possible, dramatic amounts of energy can be saved at the cost of higher water consumption. The trade-off is almost always cost effective.

It is much more difficult to generalize on ways to save in processes since water can be used in so many different ways in industrial processes. For example, water can be used to cool furnace walls, cool air compressors, wash, rinse, surface-treat, coat, test products, and cool molds and for a wide variety of other uses. Following is a discussion of some ways water and money have been saved in industrial processes:

- Use water flow restrictors in shower heads and sinks. As much as 60% savings in water and energy costs can be realized by installed flow restrictors.
- Recycle rinse water. Often rinse water can be recycled by simple filtering or treatment. In one company, $30,000 and 30×10^6 gal of water were saved annually by simply running rinse water through a sand filter and reusing it.
- Reuse cooling water. Often air compressors, small chillers, and other equipment requiring cooling are cooled with once-through cooling water (i.e., water from the tap is run through the equipment one time and dumped into the sewer). In the same room, tap water may be used as boiler makeup water. By using the cooling water as boiler makeup, significant amounts of water and energy can be saved. (The cooling water is hot, usually about 105°F.) One company found it could save about $1300 and 1×10^6 gal of water per year for *each air compressor* by reusing this water in other places or by recirculating it through a small cooling tower.
- Reduce flow rates to the minimum necessary. Usually, water flow rates are liberally set in washing, coating, or rinsing operations. By setting flow rates at minimum levels, significant water can be saved along with the energy required for pumping.

- Cover open tanks. Often heated tanks are open at the top, and floating balls, cantilevered tops, or flexible slit covers can reduce evaporation and heat loss. One plant saved about $12,000 per year in energy and water costs by covering their heated tanks.

There are many other ways to save dollars and water through water audits, but it is difficult to develop a list. Each plant and each operation are unique and require individual engineering study. The preceding discussion should serve to plant some ideas.

In terms of total water studies, water audits at manufacturing locations uncovered the following potential:

- One plant saved $77,000 annually in energy and water cost and 32×10^6 gal of water.
- Another plant saved $20,000 annually in energy and water cost and 12×10^6 gal of water.
- A rubber hose manufacturer saved $31,000 in water cost alone and 50×10^6 gal of water per year.
- A metal cylinder manufacturer saved about $20,000/year and 36×10^6 gal of water.

The potential for real savings is large. In the future, water management will be necessary as costs continue to climb and sources dry.

13.7 SUMMARY

In this chapter we analyzed alternate energy sources and water management opportunities. First, active solar systems were studied. They were found to be effective but expensive since they normally require backup and storage systems. Substantial tax credits are sometimes available. Industrial applications include recuperators, economizers, and preheating of feedstocks as well as space heating.

Then passive solar energy was examined and found to offer substantial potential in reducing energy costs for heating. Figures were developed to show that a large percentage of a plant's heating needs can be filled by passive solar energy but that backup systems are needed. Passive solar energy is cost effective today. A checklist of ideas was provided.

Then photovoltaics was examined. Photovoltaics is the direct conversion of sunlight into electricity. Present applications in industry are only for remote sites where consistency of operation is not a problem. The future is promising for reduced cost of photovoltaics.

Wind energy was examined and found, for several reasons, not to be very promising as an industrial energy source in the near future. This is especially true in congested areas where most industry is located.

Refuse-derived fuel was found to be a cost-effective energy source in many locations today. Savings occur because of reduced energy cost and reduced handling and disposal cost. Many industries find it quite profitable to use refuse as an energy source today.

A new opportunity—water management—was discussed in the last section. Water management can yield large dollar and water savings and be cost effective. Examples were provided to show that industry should be looking to water management.

REFERENCES

1. *Climatic Atlas of the United States,* GPO, Washington, D.C., 1968.
2. *Introduction to Solar Heating and Cooling Designing and Sizing,* DOE/CS–0011 UC 59A, B, C, Washington, D.C., Aug. 1978.
3. W. C. Turner, Senior ed., *Energy Management Handbook,* Wiley-Interscience, New York, 1982.
4. Michele Raymond, "Mill to Burn Rice Hulls to Make Electricity," *Energy User News,* Jan. 31, 1983.
5. Jeff Barber, "Firm Eyes Savings from Tires-To-Fuel System," *Energy User News,* Jan. 11, 1983.

BIBLIOGRAPHY

Instructions for Energy Auditors, U.S. Department of Energy, Washington, D.C., 1978.

MYERS, JOHN, *Solar,* research report, Oklahoma Industrial Energy Management Program, Oklahoma State University, Stillwater, Oklahoma, 1981.

QUESTIONS

13.1. What is a selective surface? How and why does it affect the efficiency of a solar collector?
13.2. Why would phase change materials be popular for thermal storage in solar applications where space is limited?
13.3. Describe the refuse stream of a typical university. State its probable Btu content.
13.4. What renewable energy source is most popular today? Why?
14.5. Discuss some hindrances facing wider-spread utilization of solar energy in industry.

PROBLEMS

13.1. In designing a solar thermal system for space heating, it is determined that water will be used as a storage medium. If the water temperature can vary from 80°F up to 140°F, how many gallons of water would be required to store 1×10^6 Btu?

13.2. In designing a system for photovoltaics, cells producing .5 v and 1 a are to be used. The need is for a small dc water pump drawing 12 v and 3 a. Design the necessary array but neglect any voltage-regulating or storage devices.

13.3. A once-through water cooling system exists for a 100-hp air compressor. The flow rate is 3 gal/min. Water enters the compressor at 65°F and leaves at 105°F. If water and sewage cost $1.50/$10^3$ gallons and energy costs $5.00/$10^6$ Btu, calculate the annual water savings (gallons and dollars) and annual energy savings (10^6 Btu and dollars) if the water could be used as boiler makeup water. Assume the water cools to 90°F before it can be used and flows 8760 h/year.

13.4. A large furniture plant develops 10 tons of sawdust (6000 Btu/ton) per day that is presently hauled to the landfill for disposal at a cost of $10/ton. The sawdust could be burned in a boiler to develop steam for plant use. The steam is presently supplied by a natural gas boiler operating at 78% efficiency. Natural gas costs $5.00/$10^6$ Btu. Sawdust-handling and in-process storage costs for the proposed system would be $3.00/ton. Maintenance of the equipment will cost an estimated $10,000/year. What is the net annual savings if the sawdust is burned? Assume no tax credits (although a 20% credit is probably available). The plant operates 250 days/year.

13.5. Design an energy-efficient facility (location on site, layout, building envelope, etc.) for an existing factory whose operation is familiar to you. Do not be constrained by the existing facility.

Index